MW00789847

CEREBRUM 2014

Cerebrum 2014

Emerging Ideas in Brain Science

Bill Glovin, Editor

New York

Published by Dana Press, a Division of the Dana Foundation, Incorporated

Address correspondence to:
Dana Press
505 Fifth Avenue, Sixth Floor
New York, NY 10017

Copyright 2015 by The Dana Foundation Incorporated. All rights reserved. No part of this book may be reproduced, stored in a retrieval system, or transmitted in any form or media by any means, electronic, mechanical, photocopying, recording, or otherwise, without the prior written permission of the publisher, except in the case of brief quotations embodied in critical articles or reviews.

New York, NY 10017
DANA is a federally registered trademark.

Printed in the United States of America

ISBN-13: 978-1-932594-62-1
ISSN: 1524-6205

Book design by Bruce Hanson at EGADS (egadsontheweb.com)
Cover illustration by William Hogan

CONTENTS

Foreword

By Barbara J. Culliton

Barbara J. Culliton is president of The Culliton Group, a consulting enterprise that specializes in editorial, commentary, and writing workshops in biomedicine. Culliton is a former news editor of *Science* and deputy editor of *Nature*. During her tenure at *Nature*, she founded *Nature Genetics*, *Nature Structural Biology*, and *Nature Medicine*. She is a co-winner of the George Polk Award in Journalism for her work at *Science*. In 1988, Culliton was elected to the Institute of Medicine/National Academy of Sciences, and is a former member of its governing council. She is a member of the board of the Council for the Advancement of Science Writing, was formerly the Times-Mirror Visiting Professor at the Johns Hopkins University, and a past-president of the National Association of Science Writers. She joined Celera Genomics in 1999 and founded the Genome News Network, one of the first online publications in the field. Culliton has served as an advisor or trustee of several academic institutions, including Dartmouth Medical School and Northwestern Medical School, and is currently on the board of the Institute of Human Virology at the University of Baltimore.

WHAT MAKES US HUMAN?

Is it true, as Descartes said, that "I think therefore I am," or is it the other way around: I have a brain and therefore I can think and, more important, I can create memories that help establish who I am.

What is the connection between the mind and the brain?

Why are some people extraordinarily gifted in mathematics or music or athletics or language?

Each of these philosophical questions lies behind humankind's interest in understanding itself, questions whose answers take new shape and form as neuroscience produces an astonishing abundance of insights into how the brain works.

More grounded questions are equally compelling—namely the various neurodegenerative diseases that cause dementia. As the population of Western nations ages, the incidence of Alzheimer's disease, Lewy Body dementia, Parkinson's, and other afflictions that generally, but not always, strike the elderly, is growing. The resulting human suffering and the staggering cost in health-care dollars have become the focus not only of researchers and physicians in the neurosciences but of policy-makers alike. At present, despite some therapeutic advances, these diseases are essentially untreatable in any meaningful way; nor is there any clear understanding of how to prevent them in the first place.

That said, there is no doubt that during recent decades neuroscience has been on something of a scientific roll, with new, intriguing insights appearing in the literature almost weekly. There is a feeling of being close to really knowing how neurons connect or fail to connect and what it means. With the increased availability of technologies such as functional magnetic resonance imaging (*f*MRI), it is possible to observe the human brain in action, in living people in real time. Everyone has seen the beautifully if artificially colored images of the brain as it lights up when a person is looking at art, listening to music, taking a memory test. We can also see images of areas of the brain that physically shrink in patients with certain neurological disorders, enabling scientists to link various functions to the hippocampus, the temporal lobe, the amygdala, and the prefrontal cortex.

Neuroscientists working with animal models are developing equally important data. It is possible to stimulate a single neuron in the brain of a *Drosophila*, or common fruit fly, which has a very small brain of only 100,000 neurons. Its tiny brain makes it easier to detect at the molecular level than the human brain, which has between 80 billion to 100 billion neurons. Mice and rats—mammals that make up 95 percent of all lab animals—have genetic, biological, and behavior characteristics that closely resemble those of humans, and researchers are able to replicate many symptoms of human conditions in them.

As a result of the outpouring of research and the public's understandable craving for results, President George W. Bush declared the 1990s the Decade of the Brain. A little more than a year ago, President Barack Obama caught the enthusiasm when he launched the BRAIN* Initiative, which is sometimes described as an all-out effort to map the activity of every neuron in the human brain. The Initiative carries with it a promise of $300 million in funding every year for 10 years, and is overseen by an advisory committee of some of the United States' most prominent neuroscientists. This is reassuring to researchers who lead neuroscience and technology laboratories at universities across the country; it also makes neuroscience attractive to the best and the brightest of upcoming postdocs and junior researchers. It has the same optimistic vibe as the War on Cancer, which began in 1970.

While research proceeds, advancing neuroscience literacy to the public is critical for any number of reasons. For one, reliable scientific information on issues involving the brain helps people navigate through a wide range of research and ethical issues, including false and overhyped claims, and helps consumers make more informed choices. And publications that disseminate cutting-edge neuroscience research help students appreciate the value of learning about the brain and may inspire someone who will one day become the next Nobel Prize winner.

These are among the reasons that the 2014 edition of the online journal *Cerebrum* and publications like it are so vital. This anthology of articles and book reviews contains within its pages a one-stop shopping, comprehensive snapshot (if such is possible) of what is important in neuroscience early in the 21st century. Topics include studies of how the brain makes memories and how it knows where you are, as if the brain had its own GPS system. An article on the structure of the brain describes anatomical differences between men and women. The brain processes that make us individuals and those that make some people empathetic are discussed, along with the popularity of brain games, and, of course, what happens to neurons when the brain goes off course. Technology and the brain come together in a review of paraplegia, among other subjects.

Induced pluripotent stem cells (iPSCs) in humans, which have many potential applications in health and medicine, are one of those complex subjects explained in an article by **Fred Gage** and **Carol Marchetto**. Although the promise of employing iPSCs for cell therapy is far from being realized, iPSCs are being used today in drug discovery and development. The authors describe how they were first identified, what they are, and why a growing number of researchers and clinicians believe that they may be one of the keys in treating various brain disorders.

Memory is at the heart of being human, and understanding how it is encoded is one of neuroscience's main priorities. **Paul Worley** and **Marshall Shuler** nicely summarize the complex relationships among neurons, networks in the brain, genes, proteins, receptors, and synapses—the central chemical connections that are vital to communication among the various molecules that must work together, in the right time and in the right way, to create and maintain memory. "Neurons that fire together, wire together," is one way of thinking about what is required. One characteristic symptom of Alzheimer's and other dementias is getting lost. Patients lose their sense of space and direction because cells that determine navigation are damaged.

The husband and wife team of **Edvard Moser** and **May-Britt Moser** begin an article on place cells, grid cells, and border cells by saying "the most advanced surveillance system you will ever find is built into your own brain...." Learning how grid cells and place cells communicate will contribute to understanding how the brain forms maps and how human beings use them. Knowing how these and related cells are disrupted has obvious potential for finding ways to keep them on track. How important is their research? A few months after their article appeared, the Norwegian researchers shared the Nobel Prize in Physiology or Medicine with British-American researcher John O'Keefe.

Sex differences in the human brain not only offer **Larry Cahill**'s perspective on the nature of the differences in the brain and behavior of men and women, but also what he considers to be a "counter-reaction" to such research by "anti–sex difference" investigators operating from the "deeply ingrained, implicit, false assumption that if men and women are equal, then men and women must be the same." A compelling anecdote about why the Food and Drug Administration directed the makers of Ambien to cut their recommended dosage in half to women sets the stage to ask and answer a number of important questions on why men and women are treated differently in research and medicine.

As the economic gap between rich and poor continues to widen, a topic of vital importance is addressed by **Kimberly Noble** in an article that examines socioeconomic adversity and brain development. Recent advances in neuroimaging have led to new ways of disentangling the complex interplay between genetic and environmental factors that influence structural brain development. Noble's article addresses research that investigates significant links between socioeconomic status and changes in brain structure, especially in areas related to memory, executive control, and emotion. She also tells us the reasons why addressing cognitive and socio-emotional development is so critical for children to succeed and to lead healthy, productive lives.

Richard J. Davidson's article on the neurobiology of individuality focuses on the various processes in the brain that makes us who we are. New and improved imaging evidence, a recurrent theme through so many of the articles in this anthology, is again at work in an examination of how brain circuits involved in our emotional responses are highly plastic and change with experience. Perhaps the most important part of Davidson's findings is that psychological interventions can further harness brain plasticity to promote behavior changes that increase resilience and well-being.

Concern that there may be a link between a father's age and his children's vulnerability to psychiatric problems is addressed in an article by

Brian M. D'Onofrio and **Paul Lichtenstein**. The authors talk about how they have leaned on their own study and another study to try to draw conclusions, and why so many questions on this issue still remain unanswered. But they do point out that enough empirical data exist to suggest that children born to middle-aged men are more likely than their older siblings to develop a range of mental difficulties, including bipolar disorder, autism, and schizophrenia.

Human beings, as well as animals and plants, are known to adapt to the light-dark cycle that more- or- less defines a day. Circadian rhythms control most physiological activity, usually in ways that keep organisms healthy by regulating sleeping, eating, and periods of physical or mental activity. **Paolo Sassone-Corsi** describes genetic and epigenetic research on the circadian clock which is located in a small area of the brain (the suprachiasmatic nucleus or SCN) that contains approximately 15,000 neurons, that in turn receive signals from neurons in the eye. But, contrary to previous thought, it is not the SCN alone that contains neurons that can tell time. The liver, spleen, muscle, and other body functions each have their own internal clocks. Even more recent research shows that the "clock controls a remarkable fraction of the genome...so that at least 10 percent of all expressed genes in any tissue are under circadian regulation." The author speculates, based on current data, that the circadian clock is important in host-pathogen relationships, the inflammatory response, and tumor growth.

Fresh off his well-publicized brain-to-computer demonstration during opening ceremonies at the World Cup in Brazil last summer is an article by **Miguel A. L. Nicolelis** who, for the first time in history, showed that a human subject could use a brain-controlled exoskeleton to initiate the kicking of a soccer ball. The demonstration was witnessed by 70,000 fans at the Itaquerão stadium and by an estimated 1 billion people watching on TV. Nicolelis, one of the world's pioneers in this area of research, uses references from science fiction to convince us that brain-to-computer interface will someday transform the lives of patients with serious spinal cord injury. His article traces some of the early technological breakthroughs, describes an interface he calls "Brainets," and recounts his involvement with helping create the Walk Again Project, a research consortium.

The biology of who we are and why we do what we do is at the core of an article on the underpinnings of empathy by **Peggy Mason**. Researchers such as Mason are beginning to elucidate the brain circuits that support empathy and empathetic concerns in humans using fMRI. The author is convincing when she writes that understanding empathy is the most important question we face. The subject—at the root of why politicians can't get along or how a person can brutally murder another person—is social neuroscience that may one day help us to better understand these questions.

One of the most controversial areas in neuroscience in recent years is addressed in an article that examines the effectiveness of cognitive training. **Walter R. Boot** and **Arthur F. Kramer** look at studies that measure the validity of claims made by Lumosity and other companies in the billion-dollar brain games industry. Without discounting cognitive training, they focus on problems in the research: failure to replicate findings, lack of comparative data, and misleading conclusions, to tell us why we need to look long and hard before plopping down our credit cards to invest in unproven products to ward off dementia. Playing games for their own sake is different matter.

The year's articles conclude with an essay by **Wise Young** and **Patricia Morton** on the future of scientific research. They are uniquely qualified to write on the subject, because they lead a neuroscience center that attacks chronic spinal cord injury with pluripotent stem cells, an area that holds enormous potential to advance scientific research but has been held back due to the influence of politics and religion, as well as technical challenges. Their article leans on lessons from the past, the evolving influence of technology on scientific discovery, obstacles they have personally faced, and such ideas as leaning more on collaboration than competition to generate funding, clear bureaucratic hurdles, and generate useful data.

The anthology also includes books reviews that tackle important neuroscience subjects such as autism, memory, humor, chronic traumatic encephalopathy, and prions. Who better than **Temple Grandin** to review *The Reason I Jump* by Naoki Higashida, a provocative look at the mind of a remarkable 13-year-old Japanese boy with severe autism? Grandin relates her own experience living with and studying autism to better understand the thoughts and feelings that are so sensitively translated by James Mitchell, author of *Cloud Atlas.* In **Jerome Kagan**'s review of *The Future of the Mind* by physicist and futurist Michio Kaku, Kagan leans on his own experience to explore a book that imagines a world where we will have the power to record, store, and transmit signals of brain activity, and where interchangeable thoughts and self-aware robots will be part of everyday life.

Everyone feels better when they're smiling and in **Robert Provine**'s review of *Ha: The Science of When We Laugh and Why* and *The Humor Code: A Global Search for What Makes Things Funny,* we hear him analyze why simple instincts such as laughing and yawning are important. **Philip E. Stieg,** a neuro-trauma consultant on the sidelines of NFL games, reviews *League of Denial*, one of the most talked-about sports books in recent memory. Authored by Mark Fainaru-Wada and Steve Fainaru, the book is about the impact of the sports-concussion crisis in professional football. Beyond the heart-wrenching stories used to illustrate the impact of chronic traumatic encephalopathy (CTE), as well as the moral and legal pressure and competition to advance the science, Stieg cautions us that we have much to learn about CTE.

The final review of 2014 examines Stanley B. Prusiner's *Madness and Memory*, a book that recounts the rejection and even ridicule he faced when he first said he had discovered prions, which are now known to be infectious proteins that cause neural degeneration. **Guy McKhann**, scientific consultant to the Dana Foundation, leans on his longtime relationship with Prusiner and many of the scientists and institutions that played a role in prion science to write in a personal and familiar way about the author and the importance of his research. McKhann points out that Prusiner, winner of the Nobel Prize in Physiology or Medicine, has written a book that will "enlighten and inspire you, regardless of your background." That sentiment is very much at the heart of what this anthology attempts to accomplish.

ARTICLES

1

Your Brain Under the Microscope

The Promise of Stem Cells

By Fred H. Cage, Ph.D., and Carol C. Marchetto, Ph.D.

Fred H. Gage, Ph.D., is the Adler Professor in the Laboratory of Genetics at the Salk Institute in La Jolla, CA. The Gage laboratory studies the adult central nervous system and unexpected plasticity and adaptability to environmental stimulation in mammals. In addition, he models human neurological and psychiatric disease in vitro using human pluripotent stem cells. Prior to joining Salk in 1995, Gage was a professor of neuroscience at the University of California, San Diego, where he still teaches as an adjunct. Gage served as president of the Society for Neuroscience in 2002 and as president for the International Society for Stem Cell Research in 2012. He received his Ph.D. from the Johns Hopkins University in 1976.

Carol C. Marchetto, Ph.D., is a senior staff scientist in the Laboratory of Dr. Fred Gage at the Salk Institute in La Jolla, CA. Marchetto is involved in understanding the mechanisms by which human embryonic stem cells and induced pluripotent stem cells become a fully developed functional neuron. Moreover, she is currently studying the behavior of different subtypes of human neurons in the neurological diseases such as amyotrophic lateral sclerosis (Lou Gehrig's disease) and autism spectrum disorders. Marchetto obtained her Ph.D. in genetics and microbiology in 2005 from the University of Sao Paulo, Brazil.

Editor's Note: Until recently, scientists primarily worked with two kinds of stem cells from animals and humans: embryonic stem cells and non-embryonic "somatic" or "adult" stem cells. Scientists are just now beginning to improve their understanding of a third kind: induced pluripotent stem cells. Our authors describe how they were discovered, what they are, and why a growing number of researchers and clinicians believe that they may be one of the keys in helping address various brain disorders.

WHEN THE INTERNATIONAL SOCIETY for Stem Cell Research (ISSCR) first met in Washington, D.C. in 2003, a few hundred attendees participated in the discussions. In June 2013, just 10 years later, a record 4,000 researchers from all over the world attended the society's meeting in Boston. The ISSCR now has more than 3,000 members and three affiliated indexed journals, one of which has one of the highest impact factors in the field. In addition, the number of abstracts that utilized reprogramming technology increased exponentially from basically none to more than 220 in just five years (see graph). The ISSCR's rapid growth has run parallel with an unprecedented display of general interest on the part of researchers and clinicians from different backgrounds and levels of expertise. Both trends speak directly to the potential impact of stem cell research.

We all begin our lives with one major stem cell: a fertilized egg. That one stem cell then divides and forms new cells that, in turn, also divide. Even though these cells are identical in the beginning, they become increasingly varied over time. As a result of this process, which we call cell differentiation, our cells become specialized for their locations in the body. As we develop in the womb, our cells differentiate into nerves, muscles, and so on, and the organs begin to organize and function together.

Scientists long believed that a mature or specialized cell could not "reprogram," or return to an immature state. A few researchers challenged this view, however. In 1966, John Gurdon (Wellcome Trust/CRUK Gurdon Institute, Cambridge, U.K.) was the first to show that if you removed the nucleus containing the genetic material of a fertilized frog egg (stem cell) and replaced it with the nucleus of a fully differentiated intestine cell from a tadpole, the modified egg would grow into a normal frog with the same genetic material as the original egg.[1]

Gurdon's findings were confirmed by others, including Robert Briggs and Thomas King Jr., whose earlier works showed that normal hatched tadpoles could be obtained by transplanting the nucleus of a blastula cell to the enucleated eggs of a leopard frog (*Rana pipiens*).[2] In 1997, Ian Wilmut electrofused (a technique used to fuse cells using electrical impulse) nuclei of cultured sheep adult mammary gland cells into enucleated sheep eggs and produced a single cloned sheep named Dolly.[3] These researchers sent the scientific community this message: It was now possible to reprogram adult cells to an immature state by exposing them to a yet-unknown combination of factors that were present inside enucleated eggs. These reprogrammed cells became pluripotent again, meaning they were capable of going through a new process of maturing and specializing.

Even though the pioneering researchers provided the proof of principle that reprogramming was possible, the cloning experiments they performed were very time-consuming, difficult to reproduce, extremely inefficient for mammalian cells, and ethically controversial when envisioned for human cells. In addition, an important piece of the puzzle was still missing: What made the reprogramming of adult cells possible? It was not until 2006 that Japanese researcher Shinya Yamanaka and his postdoc Kazutoshi Takahashi were able to answer this question.

The Reprogramming Pioneers

When Yamanaka presented his first reprogramming results at the 2006 ISS-CR meeting, many scientists were skeptical. Yamanaka claimed that with the addition of only four factors that are master regulators of cell pluripotency, his team could induce an adult skin cell (fibroblast) to become a pluripotent stem cell (then called an induced pluripotent stem cell, or iPS cell) within only a month. Many thought his results were too good to be true, but later that year, when his procedure was published with a description of the four factors he used for reprogramming experiments, dozens of labs around the world (including ours) tried his protocol.[4, 5] To our complete astonishment, it worked in our lab the very first time—and it worked in many other labs as well.[6,7] Yamanaka and Takahashi's research results played a major role in popularizing and disseminating stem cell research because by uncovering the basic factors and principles of the reprogramming process, they made it possible for researchers from other fields to work with pluripotent stem cells. The impact and potential of their stem cell research earned Yamanaka and Gurdon the Nobel Prize in Physiology or Medicine in 2012.

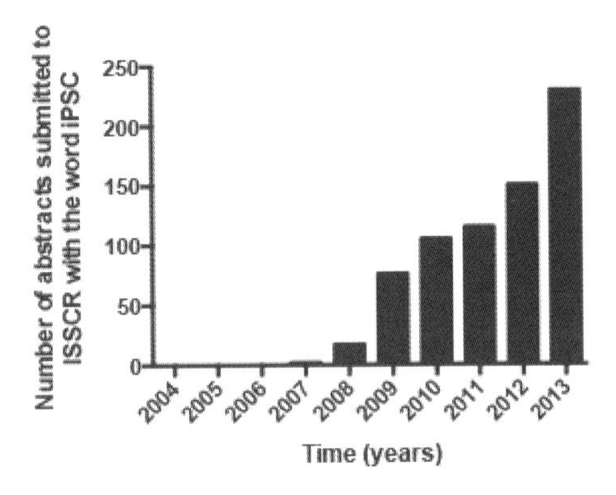

Using iPS Cells to Study Neurological Diseases

Human iPS cells, which can, in principle, form any cell in the body, could provide an attractive alternative when the traditional models for neurological diseases are inadequate.

Nearly all of our current knowledge about human neurodevelopmental and neurodegenerative diseases at the cellular level is derived from studies in postmortem brain tissues. These samples often represent the end stage of a disease and therefore are not always informative representations of a disease's developmental path. Furthermore, the pathology observed in these tissues is potentially not the authentic disease cellular phenotype. Genetically modified ("transgenic") mice provide an alternative way to reproduce human genetic forms of neurodegenerative diseases to serve as models for observation as to their developmental course in a neurotrophic phrase. However, use of these models is limited to monogenetic (the origin of diverse individuals or kinds by descent from a single ancestral individual or kind) disorders in which the specific gene mutations are known—disorders that represent the minority of neurological diseases. And in some cases, mouse transgenic technology cannot adequately model neurological disorders with defined genes because of intrinsic differences between species.

For example, mice have much less of a complex brain architecture than humans; there are a number of brain structures present in humans that are not present in rodents. This suggests a need for advancement toward human models of disease. Currently, many subtypes of disease-relevant neurons can be developed from iPS cells using a combination of manual selection and the addition of mixtures of different neurotropic factors to the culture. Differentiation protocols can provide enriched populations of particular subtypes of neurons that are relevant to specific diseases. These subtypes include dopaminergic neurons for Parkinson's disease, hippocampal and cholinergic neurons for Alzheimer's disease, motor neurons for amyotrophic lateral sclerosis (ALS, or Lou Gehrig's disease), and inhibitory interneurons for schizophrenia.[7-10]

To date, most experiments involving disease modeling for neurological diseases utilize iPS cells-derived neurons from patients with monogenetic disorders for which the gene mutation is defined and well characterized. The modeling of monogenic brain disorders has promoted rapid advancements in the field by helping to establish the basic tools for culturing functional human neurons. In addition, initial modeling research revealed meaningful neuronal phenotypes, such as differences in synaptogenesis, neuronal size and arborization complexity, and connectivity properties.[11,12] Importantly, monogenic disorder modeling presents an opportunity to perform gain-of-function and loss-of-function studies and to confirm the specificity of the neuronal phenotypes observed. In addition, studying the in vitro phenotypic consequences of the mutation in specific genes can highlight molecular mechanisms responsible for subtle alterations in the nervous system, perhaps pointing to common mechanisms for more complex, multi-gene diseases.

Nonetheless, the vast majority of neurological disorders (for example, autism spectrum disorders, schizophrenia, Parkinson's disease, Alzheimer's disease, and Lou Gehrig's disease) are complex in nature and likely multifactorial: a combination of mutations in several genes and extrinsic factors (such as influence of neighboring cells in the neuronal niche and environment) is likely involved in the disease pathology course. Recently, scientists have made successful attempts to detect a specific neuronal phenotype using sporadic neurological disease models. Hopefully there will be more advances in the near future as the technology becomes sensitive enough to detect more subtle phenotypes.[11-13]

Finding Clinically Relevant Drugs

Candidate compounds for treating central nervous system (CNS) deficiencies fail in clinical trials in more than 90 percent of cases because of poor targeting (the drug does not target the affected area of the brain efficiently), lack of efficacy, and unacceptable side effects.[14] Pluripotent stem cells derived from patients with CNS diseases offer a significant advantage, as researchers can take into consideration the patient's genetic background and the developmental course of the disease. Importantly, these stem cells allow for the generation of both genetic and sporadic forms of the disease.

Before developing a screening platform with the aim of discovering new drugs for a treatment, a consistent abnormal phenotype needs to be identified and reproduced on a large scale. Researchers are making progress in this process, and as large pharmaceutical companies move into stem cell-related drug research, more systematic progress is expected.[15-17] The best examples so far are coming from partnering between research organizations (universities and institutes) with industry and start-up companies that have scientists as advisors. A few months ago, a group from iPierian Inc. configured a high-content chemical screen using an indicator of ALS pathology in human motor neurons derived from iPS cells from patients with ALS. The group identified small molecule compounds (i.e., digoxin) that alleviated the disease-related phenotype in iPSC-derived patient neurons, thus demonstrating the feasibility of iPS cell–based disease modeling for drug screening. The general strategy for drug screening is to identify a reliable disease-related phenotype and to develop high-throughput screening platforms to test bioactive compounds (such as proteins and small molecules) that protect the patient neurons from either developing or progressing through the disease course. After rigorous testing, these screenings will likely unearth therapeutic compounds that could benefit a group of patients.

Finally, iPSCs may also be used to assess developmental as well as cell-type-specific drug toxicities. Indeed, existing commercially available human iPS-derived hepatocytes, cardiomyocytes, and neural cells may provide the basis for humanized assays to detect off-target activity and side effects of drugs in a tissue-specific manner.[18] We firmly believe that reprogramming technology can be a valuable, additional tool for screening and validating CNS compounds for pharmaceutical companies in the near future, ultimately culminating in the discovery of new therapies.

Cautionary Notes

iPS cell lines and their derived progeny bear a significant intrinsic variability, as revealed by abnormal expression of imprinted genes, differential expression profiles, and inconsistent neuronal differentiation competence.[19-21] For that reason, researchers still need to conduct comparative experiments with well-established human embryonic stem (HES) cell lines as a benchmark for complete reprogramming and ideal differentiation protocols. It is our expectation that the use of HES cell lines may decrease overtime, but studying reprogramming without them would be unconceivable at this point.

This variability can become a real hurdle for disease modeling, especially when comparing cells from patients with sporadic forms of diseases that have multifactorial etiologies. The differences observed have been generally attributed to random integration of viral vectors causing potential insertional mutagenesis, reactivation of reprogramming transgenes, and persistency of donor cell gene expression.[22] New technology that promotes the delivery of reprogramming factors in a non-integrative way is available and becoming more popular among disease modeling groups.[23,24] Reprogramming can also be achieved by using synthetic genes and small molecules, and further improvement of these methodologies will promote widespread use by the scientific community.[25] As more research groups use nonintegrative approaches, we anticipate that the iPS cell lines generated will have decreased intrinsic variability.

Identifying disease-relevant phenotypes requires researchers to compare experimental cells with "healthy" control cells. New gene-targeting technologies in iPS cells can enable more efficient and less variable rescue from monogenetic alterations. In addition, the generation of isogenic cell lines allows for more relevant controls that take into account the individual's genetic background. Examples of methods currently using iPS gene editing are zinc-finger nucleases, transcription activator-like effector nucleases, and clustered regularly interspaced short palindromic repeats.[26-29]

For sporadic cases, alternative ways to decrease variability will include using neurotypical family members as controls or including groups of patients who present common clinical histories and/or respond to drugs in a similar manner. New high-throughput genomic tools, such as genomic deep sequencing, are beginning to reveal naturally occurring genetic variation that can help us to understand the differences between cell lines. When possible, reprogramming cells from genetically identical individuals, such as monogenetic twins who are concordant or discordant for a specific neurological condition, will also help us to understand variability and to generate relevant disease hypotheses.

The Road Ahead

Reprogramming technology has opened the door for many new insights into the brain and brain-related conditions. The recapitulation of early stages of human neural development made possible by using iPS cells is an invaluable tool that can reveal the exact moment of the disease onset, thus fostering the generation of new diagnostic tools and potentially optimizing novel therapeutic interventions.

Although it has been only seven years since the introduction of somatic reprogramming technology to generate iPS cells, clinical studies that bring iPS cell–based therapy to patients are already underway. In August 2013, the Japanese Ministry of Health, Labour, and Welfare approved the first pilot clinical study using isogenic iPS cells for age-related macular degeneration (AMD). The study will be conducted mainly by the Takahashi group at the RIKEN Center for Developmental Biology in Kobe, Japan. They plan to transplant sheets of iPS cell–derived retinal cells into the subretinal space of AMD patients to rescue and restore the pigmented epithelium responsible for absorbing visual stimuli.[30] If Takahashi's study is conducted safely, it will be the first clinical demonstration of iPS cells for medical use and will undoubtedly impact the outlook regarding the safety and efficacy of iPS cell–based therapy. Advanced Cell Technology, an American biotechnology company, had applied for Federal Drug Administration approval for a less ambitious clinical trial of injecting human iPS cell–derived platelets as a potential treatment of coagulopathies. Because platelet cells lack a nucleus, scientists expect that the risks of tumors and tumor-associated immune responses will decrease. Nonetheless, the main challenge in the field remains: Much more groundwork is needed to improve understanding of the biology of reprogrammed cells and their progenies. In addition, we need to be vigilant about avoiding the dissemination of unproven applications.

Incorporation of bioengineering techniques making the use of bio scaffolds to allow for cells to grow in three dimensions will raise our level of understanding of the different brain structures and eventually begin to dissect out the birth of more complex neuronal networks. Earlier this year, an Austrian group led by Jürgen Knoblich assembled in vitro the first iPS cell–derived rudimentary brain.[31] The cerebral organoids produced by the researchers recapitulated early stages of human development (up to approximately nine weeks of pregnancy) and modeled for microcephaly, a neurological condition that is not efficiently modeled in rodents. More refinement of the technique will be required in order to maintain the cells as organoids or tissue in a viable and stable state for longer periods; nevertheless, the tissue-engineering approach is a very promising and powerful tool for understanding various aspects of human brain development. Neuroscientists in the past could not have predicted a scenario in which patient-derived, live functional neurons would be readily available for research, and researchers in the future will not be able to imagine a scenario without it.

2

Solving the Mystery of Memory

By Paul Worley, M.D., and Marshall G. Hussain Shuler, Ph.D.

Paul Worley, M.D., is a professor in the Department of Neuroscience at the Johns Hopkins School of Medicine. He joined the department as an assistant professor in 1988 and became a professor in 1999. His laboratory examines the molecular basis of learning and memory. The lab cloned a set of immediate early genes that are rapidly transcribed in neurons involved in information processing and essential for long-term memory. Worley received his medical degree from the University of Pittsburgh in 1980 and his B.A. and M.A. in chemistry from the Johns Hopkins University.

Marshall G. Hussain Shuler, Ph.D., joined the Department of Neuroscience at the Johns Hopkins School of Medicine as an assistant professor in 2008 after completing his post-doctoral research with Dr. Mark Bear at the Picower Institute for Learning Memory at MIT as a Howard Hughes fellow. He received a Ph.D. in neurobiology from Duke University in 2001, working in the laboratory of Miguel Nicolelis, having received a National Research Service Award fellowship. He received a B.A. with distinction in neuroscience from the University of Virginia.

Editor's Note: The word "memory" is derived from the ancient Greek myth of Mnemosyne, the mother of the Muses, who was "said to know everything, past, present, and future." Memory is essential to our existence, and one of neuroscience's primary missions is to understand how the brain processes memory and to improve treatments for Alzheimer's disease, traumatic brain injury, drug addiction, and the many other afflictions associated with disrupted memory. Our article traces scientists' progress in understanding memory over the last 15 years.

IN THE 1980S, researchers developed experimental tools that showed enormous potential in helping us move toward a molecular understanding of memory, an enduring and ongoing challenge. This is where my own (P.W.) career in basic science research really began. As I read papers by leaders in the field, the goal and experimental approach seemed clear: To identify the molecules of memory and determine how they worked. All these years later, we have made great strides that can be appreciated in the story of a single protein in the brain that plays a key role in strengthening or weakening processes involved in memory.

There are several elements other than this protein that factor in (neurons, networks, genes, receptors) and, in explaining the complexities of how all the pieces fit together, it is best to start with synapses, which are chemical connections between neurons that are also thought to play a central role. Synapses contain both a presynaptic element that releases neurotransmitters and a postsynaptic element that includes receptors for neurotransmitters. Neurotransmitters are released in bursts and bind to receptors, which then become activated. Receptors function with other proteins to convert a signal from the activated receptor into changes in properties of the postsynaptic cell, in a process termed signal transduction. Synapses can be strong if they produce large changes in the postsynaptic cell, or weak if they produce little or no change. The strength of synapses can be modified by use, and this activity-dependent change in synaptic strength is a key to understanding memory.

The perception among neuroscientists is that specific synapses are activated as specific neurons are activated, and this results in enhanced connections between certain neurons, but not between others. If one of the resulting networks of interconnected neurons is activated, others are also likely to be activated, and this enhanced connectivity encodes information. Think of it as "neurons that fire together, wire together." The mystery behind it all is how specific synapses are made to become weaker or stronger, and how these changes in synaptic strength are maintained for long periods of time.

Many diseases of the brain are associated with disrupted memory. This association is especially strong for Alzheimer's disease, traumatic brain injury, and stroke. But concepts of memory are also central to understanding substance addiction, in which the drug usurps brain mechanisms that normally reward certain behaviors.[1] Indeed, drug addiction provides a window to an important frontier of memory research that involves understanding the relationship between reward and learning (reinforcement learning theory). Many of the following insights emerged from studies of cocaine addiction.

A Small but Vital Protein

More than 50 years ago, conventional studies showed that the synthesis of ribonucleic acid (RNA) and their translation into new proteins are required for animals to establish long-term memory.[2] In these studies, injection of chemicals that blocked RNA or protein synthesis blocked long-term memory. The finding might not seem surprising since neurons are made of proteins. However, the key observation is that the processes of RNA synthesis and subsequent expression of protein are required only during a brief window of time immediately following a significant experience. Various experiments determined that this time window is about three hours. This finding indicated that mRNAs and the proteins they generate within this time window make or allow memory to happen.[3] However, scientists still had to identify the mRNAs and determine how the proteins they encode function to facilitate memory.

The experimental tools developed in the 1980s allowed researchers to identify mRNAs that are rapidly induced in cells.[4,5] I (P.W.) was fortunate to be in the right place to develop animal models and molecular techniques to identify a set of genes that were rapidly increased for one to three hours during the time most critical for memory. These genes are termed cellular immediate early genes (IEGs), a term that recognizes their rapid and transient increase in cells. One of these IEGs encodes a protein named Homer1a.[6] This small protein is present in the cytoplasm of neurons and enriched in the neurons' dendrites, which receive electrochemical signals from other neurons. With help from Dr. Daniel Leahy of Johns Hopkins, we were able to visualize the molecular structure of Homer1a and determine that it encodes a single module for binding other proteins.[7] An obvious question was: How does this small protein, Homer1a, contribute to memory? Studies over the past 15 years support a model in which the Homer1a protein binds to a neurotransmitter receptor located at the synapse and changes its properties in a way that can enhance active synapses and suppress inactive synapses. This complex process reveals several concepts important to the study of memory, which we will describe more fully.

For the purpose of description, imagine a single neuron in an animal's hippocampus—a brain region that is important for memory. The neuron fires electrical discharges as the animal experiences its environment. This firing may happen as a consequence of finding food or some other event that is important for the animal, and it results from the activation of synapses onto the activated neuron. Electrical activation of the neuron induces transcription of the genetic code from the Homer1 gene to generate Homer1a mRNA. The Homer1a protein that is produced then moves from the nerve cell body to its dendrites, which are rich with synapses. This process takes about 45 minutes. Synaptic activity that induces generation of the Homer1a protein also produces long-lasting changes at the active synapse. Neurobiologists conceptualize these changes as a "tag" that makes them different from synapses that are not active.[8] When the Homer1a protein arrives at synapses, it binds to a neurotransmitter receptor for the excitatory neurotransmitter glutamate, which is involved in learning and memory. The receptor is called metabotropic glutamate receptor type 5 (mGluR5). Homer 1a causes changes in synaptic function according to whether the synapse was tagged or not. If the synapse was not tagged, Homer1a causes weakening of the synapse. Inversely, if Homer1a binds to mGluR5 at a synapse that *was* tagged, it strengthens the synapse. The processes by which Homer1a weakens or strengthens synapses are very different and complex, as we describe in our next section.

A Protein's Role at Synapses

Homer1a reduces the strength of non-tagged synapses. The process begins as Homer1a protein increases in neurons that are strongly activated. In natural conditions, neurons are activated by synaptic input, but only a small percentage of the total number of synapses are activated. The majority of the synapses are quiet. At these quiet synapses, Homer1a protein binds to the mGluR5 receptor and activates this receptor by a process that does not require the neurotransmitter glutamate to be involved.[12] This is unusual for a neurotransmitter receptor, but is a phenomena that is part of a well-established pharmacology for mGluR5. The Homer1a-activated mGluR5 receptor then produces changes in signaling that results in the removal of a different class of glutamate receptor, called GluA2, from the synapse.[11] GluA2 is one of the major receptors that determine synaptic strength. Removal of GluA2, therefore, results in a persistent weakening of synapses. This can be considered a homeostatic process that prevents persistent increases of neuronal activity, which may be important to prevent damage to the neuron.

Homer1a can also strengthen synapses. This strengthening action occurs specifically at synapses that were activated during the process that increased neuronal activity and Homer1a expression, and occurs at the same time as a weakening of inactive synapses occurs. The process of synaptic strengthening is dependent upon changes of the mGluR5 receptor occurring at individual activated synapses. Dr. Anne Young's laboratory at the Massachusetts General Institute for Neurodegenerative Disease reported that the mGluR5 receptor is phosphorylated (undergoes a chemical change) at the same position, whereas Homer1a protein binds to it.[9] Phosphorylation often changes the way proteins function, and the fact that this occurred at the Homer binding site suggested that it might change the mGluR receptor in some

Graphical Abstract: Proposed model of mGluR5 signaling and cocaine plasticity in motor sensitization.

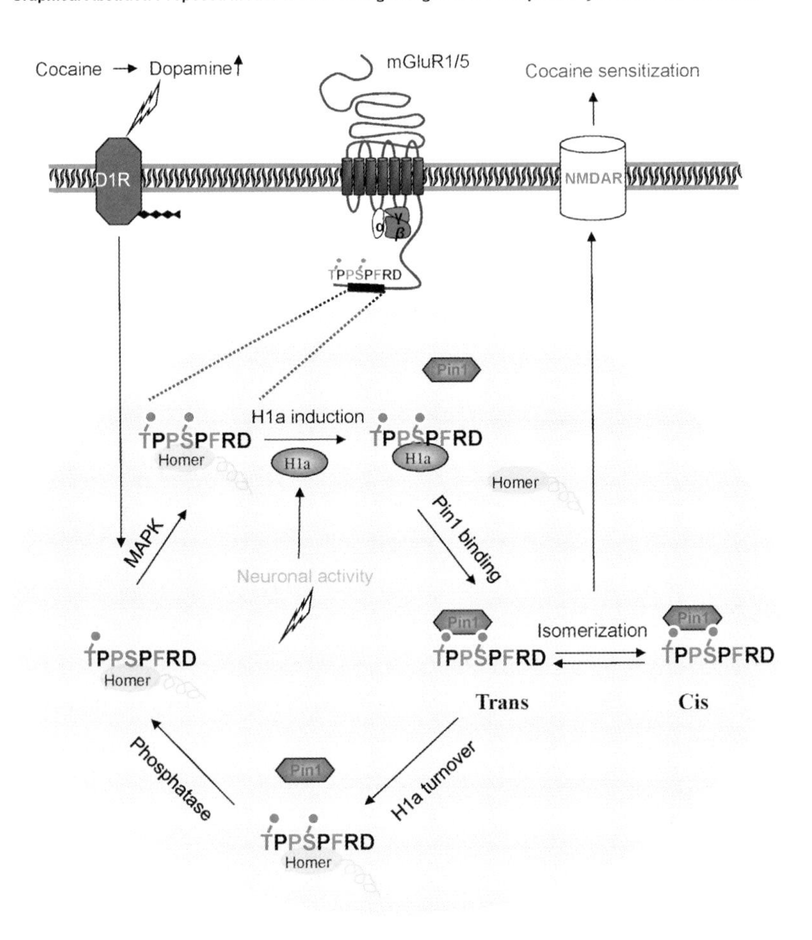

The abstract illustrates a role for mGluR5 in coupling dopamine receptor D1R activation to cocaine addiction. Critical events include phosphorylation of the Homer binding site in mGluR5, which creates a potential binding site for Pin1. The immediate early gene Homer1a is required for Pin1 binding, isomerization and enhanced activation of NMDA receptor. Enhanced NMDA receptor function mediates cocaine sensitization, an aspect of addiction. (Figure credit: Dr. Jia Hua Hu)

important, though unknown way. Phosphorylation of mGluR5 is mediated by kinases (enzymes) that target the amino acids (building blocks of protein) serine or threonine that are adjacent to the amino acid proline (enzyme class is termed proline-directed kinase). When the receptor for the neurotransmitter dopamine (which is involved in the brain's reward system) is activated, the proline-directed kinase (termed mitogen-activated protein kinase MAPK; also termed ERK), increases phosphorylation of the mGluR5 receptor.[10] This process is important for understanding memory formation, since dopamine is a neurotransmitter that can change neuronal properties to enhance brain activity and memory (termed "neuromodulator"), and is released by special neurons in the brain in association with pleasurable events.

Addictive drugs such as cocaine and amphetamine also increase dopamine in the brain. One consequence of mGluR5 receptor phosphorylation is that the Homer1a protein binds more strongly. A second consequence is that it "creates" a binding site for an enzyme termed Pin1.[10] Pin1 is a prolyl isomerase that accelerates the rotation of the amino acid backbone of the phosphorylated serine-to-proline bond and changes the three-dimensional structure of the protein. Pin1 binding requires phosphorylation of mGluR5, as well as Homer1a. Binding and isomerization are biophysical events that can be visualized at the level of single atoms using purified proteins and special techniques. This analysis reveals the relationship between binding of Pin1 and the motion of mGluR5. Together, these observations help us create an integrated model in which dopamine release in the brain increases rotational motion of the region of mGluR5 that connects with other proteins. In this way, dopamine changes the ability of mGluR5 to produce signal transduction in the postsynaptic neuron. But then what?

A Tie to Drug Addiction

Electrical recordings revealed that mGluR5 activation results in opening ion channels in the neuronal cell membrane. This opening results in a net movement of positive charge (sodium and calcium ions) into the neuron (inward current). Ion movement creates an electrical current that can be detected and recorded. This particular current is created by opening a third type of glutamate receptor termed the N-methyl-D-aspartate (NMDA) receptor.[10] Details of this process are unknown, but we imagine that increased rotational motion of mGluR5 created by Pin1 causes the NMDA ion channel to open. The NMDA receptor is critical in this pathway, as it is known to produce changes in the postsynaptic neuron that increase the strength of synapses. Importantly, the ability of mGluR5 to activate the NMDA channel increases when mGluR5 is phosphorylated and both Homer1a and Pin1 are functional at the synapse. Mice whose mGluR5 gene has been mutated so that mGluR5 cannot be phosphorylated show altered responses to cocaine. Several experimental manipulations, including interruption of Homer1a induction, interruption of phosphorylation of mGluR5, or the reduction of Pin1 activity, all similarly reduce a cocaine-induced behavior linked to addiction.[10] This makes us believe that the biophysical events that mediate Homer1a-Pin1-mGluR5 potentiation at individual synapses are causal for cocaine addiction.

Enhancing Memory

How does the biochemical model of Homer1a-dependent synaptic modification work to enhance memory in the brains of animals? The role of Homer1a in memory is easiest to conceptualize in the hippocampus, an area for which we know a great deal about IEG expression and neuronal activity linked to memory. Recordings of hippocampal neurons, from conscious, behaving rats reveal that specific neurons are consistently activated in association with an animal's understanding of its position in space. These neurons are called place cells. Dr. Bruce McNaughton, then of University of Arizona, created electrodes (termed tetrodes) that could record from multiple neurons simultaneously,[13] and monitored activity during both active exploration and subsequent periods of rest in a rat's home cage. Dr. McNaughton and his then-postdoc Dr. Matthew Wilson (currently of MIT) discovered that patterns of neuronal activity that occurred during active exploration were "replayed" later while the rat was resting.[14] Subsequent studies by Drs. Wilson, David Foster, and others have shown that replay occurs during both rest and active exploration.[15] Moreover, aspects of the replay can be used to predict subsequent behaviors, as if we are listening in as neurons plan behaviors, in addition to reviewing significant daily events.[16]

Place cells and replay are also evident in the pattern of IEG expression in the hippocampus. Dr. John Guzowski, working with Drs. Carol Barnes and McNaughton, noted that mRNA transcripts of the IEGs Arc and Homer1a could be detected in place cells using an in situ hybridization method that distinguishes between mRNA that is present in the nucleus and mRNA that is present in the cytoplasm of neurons.[17,18,19] As predicted from in vivo recording data in the rat, specific neurons of the hippocampus showed place-specific activation that was consistent across time for the same place. Moreover, IEG activation occurred after the rat had returned to its home cage in the same neurons that were active during behavioral exploration.[20,21] This late re-expression of IEG expression is consistent with replay (as originally described). Interestingly, the size of the population of neurons that re-express IEGs is smaller than the original place-cell population, as if the process that generates replay selects a subpopulation of place cells. This may represent selection of the rat's most important events for long-term storage.

Reinforcement Learning

Advances in understanding memory result from many different approaches. For example, methods used to identify brain IEGs have utilized viruses that naturally infect bacteria to store genetic information, and yeast cells whose growth can tell us about protein interactions. Investigators also use theoretical and computational approaches that are based on logic and systems or psychophysical data (the area of research that captured the imagination of author Marshall Shuler). In most cases, it is difficult to relate the empirical and the theoretical directly. But we can speculate on how certain theoretical properties of a learning model correspond to specific events or molecules from empirical data. We begin with a theoretical description of memory and definitions of critical events that must occur for memory storage.

We can define learning as relating outcomes based on behavioral import to preceding neural activity that engendered the goal-seeking behavior. This process occurs through the modification of synaptic strengths so as to make a given behavior more or less likely. Knowing which of the many synapses in a network (or indeed, in the brain) to modify poses a challenge, for the intricate pattern of activity preceding any outcome spans a spatially diffuse and vast array of synapses active at different moments in the past, not all of which are pertinent to the behavior to be learned. The attribution of "credit" to synapses that were causative in generating a behavior—the so-called spatiotemporal credit assignment problem—is therefore a central problem in neuroscience, addressed in reinforcement learning theory. In reinforcement learning, the relating of neural activity to future outcome requires an interaction between a signal conveying the success or failure of the behavior—the reinforcement signal—with a memory of which synapses were involved—referred to as an eligibility trace—in the stimulus-action response. The nature of the interaction between eligibility traces (that serve as memories of synapses' activity histories) and reinforcement signals is described as a synaptic learning rule. Such a rule, by predicating changes in synaptic strength on the presence of a reinforcement signal and eligibility traces, can resolve the credit assignment problem.[22] However, biological instantiations of the reinforcement signal, and particularly the putative synaptic eligibility trace, are unknown, as is the synaptic learning rule that governs their proposed interaction.

The processes theorized in reinforcement learning finds support in the molecular biology of the synapse. Imagine a neuron that communicates with many different neurons, and that communication with one particular neuron is important for behaviors that are rewarded, while other synapses and behaviors are not rewarded. In this scheme, synaptic interactions that cause neuronal activation (spiking) are identifiable by increased NMDA receptor–mediated currents. This transient activation occurs similarly at synapses that are rewarded or not rewarded. For those behaviors and associated synapses that are ultimately (later) rewarded, the reinforcement signal—here proposed to be the neurotransmitter dopamine—must interact in some way with the preceding distribution of activated synapses. One possibility is that the biochemical synergy between NMDA receptor activation and dopamine receptor activation for triggering of the mGluR5 kinase MAPK described above may persist even if NMDA receptor is activated for some brief time (a fraction of a second to several seconds) before arrival of the dopamine receptor signal. In this scenario, MAPK would be preferentially activated at synapses that are associated with rewarded behaviors and create an eligibility trace in the form of phosphorylated mGluR5. Phosphorylated mGluR5 at specific synapses awaits the arrival of Homer1a from the nuclear transcription response. Together with Pin1, they cause a sustained increase of synaptic strength. In this manner, credit is assigned to a synapse in accordance to the degree in which its activity is predictive of future reinforcement, and is converted, albeit at a delay, to a change in synaptic strength as plasticity proteins subsequently arrive. NMDA-dependent signals at synapses that were activated with behaviors that were not rewarded will decay prior to arrival of Homer1a and be consequently weakened. This model rationalizes the time (45 minutes) requirement for IEG expression as necessary to permit decay of non-rewarded synapses. By this speculative mechanism, synapses linked to rewarded behaviors are strengthened relative to other synapses, and this creates the substrate for long-lasting memories.

Conclusion

Important advances come with the integration of divergent ways of thinking. IEGs stand at the boundary between synaptic biology and systems physiology, and establish a framework for instantiation of reinforcement learning theory. Their selective expression in neurons that are engaged in information processing and storage affords a means to visualize and even to manipulate individual neurons or networks of neurons in order to understand their contribution to memory. IEG action in neurons reveals ways in which synapses are modulated as neurons and networks store information. Computational and theoretical models help place these molecular ideas into action and focus attention on the biochemical events that are most consequential, with the ultimate goal of understanding memory and preventing diseases that disrupt memory.

3

Mapping Your Every Move

By Edvard Moser, Ph.D., and May-Britt Moser, Ph.D.

Edvard Moser, Ph.D., is director and a founder of the Kavli Institute for Systems Neuroscience in Norway, as well as co-director of the Centre for Neural Computation at the Norwegian University of Science and Technology. His work, conducted with May-Britt Moser as a long-term collaborator, includes the discovery of grid cells in the entorhinal cortex, which provides the first clues to a neural mechanism for the metric of spatial mapping. In 2013, he, his wife, and mentor John O'Keefe were awarded Columbia University's Louisa Gross Horwitz Prize for discoveries that have illuminated how the brain calculates location and navigation. He is a member of the Royal Norwegian Society of Sciences and Letters, and the Norwegian Academy of Science and Letters. He was awarded a Ph.D. in neurophysiology from the University of Oslo in 1995.

May-Britt Moser, Ph.D., is co-director and a founder of the Kavli Institute for Systems Neuroscience in Norway, as well as director of the Centre for Neural Computation at the Norwegian University of Science and Technology. Her work, conducted with Edvard Moser as a long-term collaborator, includes the discovery of grid cells in the entorhinal cortex, which provides the first clues to a neural mechanism for the metric of spatial mapping. In 2013, she, her husband, and mentor John O'Keefe were awarded Columbia University's Louisa Gross Horwitz Prize for discoveries that have illuminated how the brain calculates location and navigation. She is a member of the Royal Norwegian Society of Sciences and Letters, and the Norwegian Academy of Science and Letters. She was awarded a Ph.D. in neurophysiology from the University of Oslo in 1995.

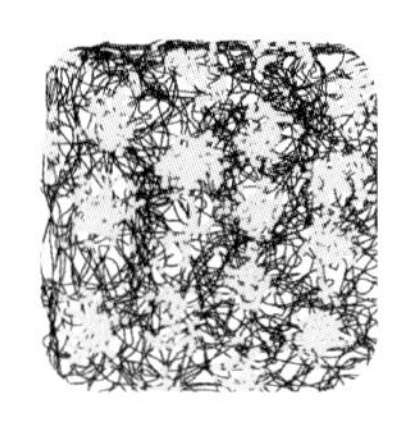

Editor's Note: In 2005, our authors discovered grid cells, which are types of neurons that are central to how the brain calculates location and navigation. Since that time, they have worked to learn how grid cells communicate with other types of neurons—place cells, border cells, and head direction cells—to affect spatial awareness, memory, and decision-making. Because the entorhinal cortex, which contains the grid-cell navigation system, is often damaged in the early stages of Alzheimer's disease, future research to better understand how cognitive ability and memory are lost has great potential significance for the treatment of Alzheimer's and other neurological disorders.

THE MOST ADVANCED SURVEILLANCE SYSTEM you will ever find is built into your own brain and nurtured by evolution. It comes equipped with a coding system that stockpiles and maps your lifetime of events in high definition. Through new research tools and insights, scientists are gradually coming to understand this coding system and its intrinsic mathematical principles.

Researchers have long known that different kinds of neurons play different roles in the brain, but only in the past few decades have scientists had access to the imaging and measurement tools they need to see how different neurons react when the brain is challenged with different tasks. We have focused on understanding how the brain helps us navigate in the environment, both because it is intrinsically interesting and because it turns out that navigation—finding our way—is linked to the way we store memories.

We now know that this coding system works like your own air traffic controller—monitoring every movement you make, knowing every step you ever made, and creating links to every event and experience you have had. Essentially, while your brain is making mental maps to help you navigate, it is also overlaying memories—experiences, smells—onto those maps.

From Map to Memories

This ability of the brain to overlay recollections creates a cognitive map—a multilayered collection of memories—rather than a mere cartographic map. It also means that learning how the brain computes navigation is a step toward understanding how networks are built up in the cerebral cortex, the part of the brain that is responsible for imagination, reasoning, and planning—thought processes that make us human.

Further insight into how the brain builds networks in the cerebral cortex can potentially lead to interventions that spare millions of people from the debilitating effects of brain disorders and diseases. The economic consequences are already significant. One study attributes an estimated 35 percent of all burden of disease in Europe to brain disorders.[1] Another study put the total cost of treatment for brain disorders in 2010 at roughly $1.09 trillion.[2] Partly due to economic forecasts, the Obama administration has recognized the importance of funding basic neuroscience research by establishing the National Institutes of Health's Brain Research through Advancing Innovative Neurotechnologies (BRAIN) Initiative.

All of these factors underscore why developing insights into the detailed workings of the brain is pivotal for both preventing and treating disorders of memory, and why focusing on the workings of the mammalian spatial-navigation system is so crucial. While what we do is basic research, our work nevertheless examines the very same system that collapses in the case of dementias such as Alzheimer's disease. Every new piece of knowledge that researchers gather contributes to understanding the puzzle posed by the brain. And we are only just beginning to see the bigger picture.

Encoding Experience on a Map

For a long time psychologists have studied how animals move in and relate to space as a way to understand the larger rules governing how and why we do what we do. Initially, most scholars thought that behavior was simply a matter of stimuli triggering responses. But in 1948, cognitive physiologist Edward C. Tolman suggested a new way to view behavior. The brains of humans and other animals, Tolman said, have a kind of map of their spatial environment, and they encode experience on top of that map.[3] This idea led to the introduction of the cognitive map.

Tolman's idea was debated but not fully accepted until 1971, when John O'Keefe and John Dostrovsky discovered place cells.[4] Place cells, which fire when an animal is in a specific place, are located in the hippocampus, a paired structure deep inside the brain underneath the cerebral cortex. In experiments, these cells "fired" whenever a rat was in a certain place in its local environment—an indication that there was suddenly something in the brain that actually looked like a map. This finding also helped demonstrate that humans and other animals could make mental maps rather than simply relying on landmarks.

In 1978, O'Keefe and Lynn Nadel took this one step further, proposing that place cells provide animals with a dynamic, continuously updated representation of space and the animal's position in space.[5] Tolman, it seemed, was right after all.

Discovering Grid Cells

The discoveries of the 1970s gave scientists more clues about what to look for and where to find it. So they looked—and found. One key discovery was head direction cells, which fire when animals face in a certain direction, regardless of the animal's position.[6,7]

In 2005, our laboratory discovered another kind of cell, called the grid cell, which is located in the entorhinal cortex, right next to the hippocampus.[8] As its name suggests, grid cells create a regular, triangular grid by firing when an animal passes over equally spaced locations. The grid looks very much like the pattern on a Chinese checkers board.

Three years later, our lab and another lab simultaneously reported the existence of yet another type of cell, called a border cell, which fires when an animal is near its environment's border, such as a wall or an edge.[9,10]

The collective significance of these findings is that the reactions of the neurons can be matched to what is found in the external world. It is still too difficult to trace other types of complex thinking to their sensory origins.

Where information is combined across sensory systems, the firing patterns of the neurons involved are too diffuse for us to detect patterns and relationships to what is happening in the external world.

HM's Legacy

We know the hippocampus is critical in forming memories, in part because of the unfortunate experiences of an American patient known as HM, who had surgery in 1953 to remove most of his hippocampus as a cure for his extreme epilepsy. The surgery succeeded on one level by reducing his epileptic seizures, but it left him unable to make new memories.

HM is not the only patient whose experiences have illuminated the workings of memory. Many others who suffered injuries to the hippocampus have helped underscore the important connection between this part of the brain and memory formation. We know that one of the first symptoms of Alzheimer's disease is that patients get lost—and that the first place where a patient's brain cells begin to die is in the entorhinal cortex.

Essentially, if you have lesions in these areas of the brain, you lose your ability to find your way—and your ability to recall all other types of memories. Memory is deeply and physically connected to our perception and encoding of space. Thus, a detailed understanding of this region of the brain and the operations of its neurons may have the additional benefit of illuminating the mechanisms behind Alzheimer's disease and related dementias.

New Technologies

As young researchers, what we most wanted to understand was behavior, as well as the origins of complex psychological functions. It's a question that will take many lifetimes to answer. So by focusing on something more accessible, such as the way space is represented in the brain, we can begin to understand how the brain computes itself, and how external inputs from the senses get into the primary sensory cortex.

Finding these cells required us to use microelectrodes, tiny wires that are thinner than a human hair. They must be correctly placed close to individual neurons in the brains of rats to allow the firing of the neurons to be recorded.

A rat's brain is the size of a grape. Inside there are about 200 million neurons, each of which has direct contact with approximately 10,000 other neurons. Inside each side of the rat's grape-size brain are areas that are smaller than a grape seed—collectively, the hippocampus—where memory and the sense of location reside. This is also where we find the place cells—neurons that respond to specific places. But from which cells do these place cells get information?

The answer is to look "upstream" of the hippocampus, to the entorhinal cortex, which feeds information to the hippocampus.

"Listening In" to Neurons

Microelectrodes allow us to listen in on the electrical activity of the cells inside the entorhinal cortex. We have advanced this technique to a level at which we can listen to several hundred cells inside a single rat's entorhinal cortex. Listening to many hundreds of cells has allowed us to discover that the brain has a number of modules dedicated to self-location.[11] Each module contains its own internal, GPS-like mapping system that keeps track of movement, as well as other characteristics that distinguish it from other modules.

Different modules react differently to changes in the environment. For instance, some scale the brain's inner map to the animal's surroundings, while others do not. And the modules operate independently in several ways. The brain can use this independence to create new and varied combinations—a very useful tool for memory formation.

This finding suggests that the ability to make a mental map of the environment arose very early in evolution. All species need to navigate, so that some types of memory may have arisen from brain systems that were initially developed for the brain's sense of location.

The grid cells in each of the brain's modules send signals to the place cells in the hippocampus. The combined effect of this grid cell activity creates an activity field in the hippocampus, the place field. This signaling, in a way, is the next step in the progression of signals in the brain. When the environment changes, the different grid modules react differently to the change—firing at new positions in the environment, and the linear summation activates different place cells in the hippocampus.

In practice, this means that the grid cells send a different combinatorial code into the hippocampus in response to the slightest change in the environment. So every tiny change results in a new combination of active cells—cell ensembles that can be used to encode a new memory, and that, with input from the environment, become what we call memories.

Neurons Talking

Recent advances in technology have given us opportunities that we could barely dream of only a few years ago. One is the ability to create detailed functional maps that show which neurons talk to each other. We are particularly interested in how grid cells and place cells communicate. The answer to this question will allow us to understand how the deepest parts of the brain are wired together.

When neurons send signals to each other, they share many similarities with electric cables. They send an electric current in one direction—from the "body" of the neuron and down a long arm, called the axon, which extends to the branched arms, or dendrites, of the nerve cell next in line. Brain cells thus get their small electric signals from a whole series of such connections.

A recent technique in our lab involves using a highly modified adeno-associated virus (AAV) as a biological transport system within neurons to better understand which neurons talk to place cells in the hippocampus. The virus is modified so that it can enter specific neurons and travel upstream through the axon and into the dendrites. We attach a light-sensitive gene to this viral transportation system. This gene integrates itself into the neuron's DNA and makes the neuron light sensitive. Normally, of course, the neuron is tucked away in the deepest recesses of the brain, in the dark. So this process allows us to install the equivalent of a light switch in a neuronal network.

We used this technique to insert light switches into place cells. Then we inserted optical fibers into a rat's brain, which enabled us to transmit light to the place cells that had light switches in them. We also implanted thin microelectrodes between the cells to detect the signals sent through the axons every time the light from the optical fiber was turned on. This allowed us to see exactly how the cell-to-cell communication was wired and to map small and large networks within the navigational computation system of the brain.

Mysteries Remain

When we put together all the information, we saw that there is a whole range of differently specialized cells that together provide place cells with their information.[12] The brain's GPS—its sense of place—is created by signals from place cells to head direction cells, border cells, grid cells, and cells that have no known function in creating location points. Place cells not only receive information about a rat's surroundings and landmarks, but also continuously update their own movement—an activity that is actually independent of sensory input.

We were surprised to find that cells that have no role in our sense of location actually send signals to place cells, because until now, the specific kinds of brain cells found to be involved in navigation—place cells, head direction cells, and grid cells—all have specific jobs. What is the role of the cells that are not actually part of the sense of direction? They send signals to place cells, but what do they actually do? This remains a mystery.

We also wonder how the cells in the hippocampus are able to sort out the various signals they receive. Do they "listen" to all of the cells equally effectively all the time, or are there some cells that get more time than others to "talk" to place cells?

Speed Cells and Decision-Making

It is easy to forget, as we move effortlessly from home to job, or from job to supermarket to home, the enormous number of processes and steps that make up our ability to navigate. We are now working our way through different aspects of the brain's navigational system to better understand how all these pieces fit together.

At the moment we are studying what we have dubbed speed cells—cells that react exclusively to the speed of an animal's movement—and how these types of cells factor in to the navigational equation.

We're also looking at decision-making. As an animal moves through a labyrinth, it must choose which way to go or what turn to make next. The neurons involved in this decision-making can be found in the prefrontal cortex, which connects to the hippocampus via a small nucleus in the thalamus.

Slowly but surely, we and other researchers are expanding our understanding of other parts of the brain to figure out how everything is connected. And because everything is connected, we are hopeful that as we and others make ever more detailed maps of neural networks, we become more and more likely to find clues that will help prevent and cure brain diseases in the future.

4

Equal ≠ The Same

Sex Differences in the Human Brain

By Larry Cahill, Ph.D.

Larry Cahill, Ph.D., is a professor in the Department of Neurobiology and Behavior at the University of California, Irvine. He first became interested in brain and memory as an undergraduate at Northwestern University. After working on memory-enhancing drugs at G.D. Searle & Company in Illinois for two years, he earned his Ph.D. in neuroscience from the University of California, Irvine, in 1990. Following postdoctoral research in Germany, he returned to UC Irvine to extend his research to studies of human subjects, which in turn led to his discoveries about sex influences on emotional memory, and to his current general interest in the profoundly important issue of sex influences on brain function. His work has been highlighted in the *New York Times*, *London Times*, *Frankfurter Allgemeine Zeitung*, and on PBS, CNN, and *60 Minutes*.

Editor's Note: While advances in brain imaging confirm that men and women think in their own way and that their brains are different, the biomedical community mainly uses male animals as testing subjects with the assumption that sex differences in the brain hardly matter. This article highlights some of the thinking and research that invalidates that assumption.

EARLY IN 2013, THE FOOD AND DRUG ADMINISTRATION (FDA) ordered the makers of the well-known sleep aid Ambien (zolpidem) to cut their recommended dose in half—but only for women. In essence, the FDA was acknowledging that despite extensive testing prior to the drug's release on the market, millions of women had been overdosing on Ambien for 20 years. On February 9, 2014, CBS's *60 Minutes* highlighted this fact—and sex differences in general—by powerfully asking two questions: Why did this happen, and are men and women treated equally in research and medicine?[1]

The answer to the first question is that the biomedical community has long operated on what is increasingly being viewed as a false assumption: that biological sex matters little, if at all, in most areas of medicine. The answer to the second question is no, today's biomedical research establishment is not treating men and women equally. What are some of the key reasons for the biomedical community's false assumption, and why is this situation now finally changing? What are some of the seemingly endless controversies about sex differences in the brain generated by "anti–sex difference" investigators? And what lies at the root of the resistance to sex differences research in the human brain?

Why Sex Didn't Matter

For a long time, for most aspects of brain function, sex influences hardly mattered to the neuroscience mainstream. The only sex differences that concerned most neuroscientists involved brain regions (primarily a deep-brain structure called the hypothalamus) that regulate both sex hormones and sexual behaviors.[2] Neuroscientists almost completely ignored possible sex influences on other areas of the brain, assuming that the sexes shared anything that was fundamental when it came to brain function. Conversely, the neuroscience mainstream viewed any apparent sex differences in the brain as not fundamental—something to be understood after they grasped the fundamental facts. By this logic, it was not a problem to study males almost exclusively, since doing so supposedly allowed researchers to understand all that was fundamental in females without having to consider the complicating aspects of female hormones. To this day, neuroscientists overwhelmingly study only male animals.[3]

To make matters worse, studying sex differences in the brain was for a long time distasteful to large swaths of academia.[4] Regarding sex differences research, Gloria Steinem once said that it's "anti-American, crazy thinking to do this kind of research."[5] Indeed, in about the year 2000, senior colleagues strongly advised me against studying sex differences because it would "kill" my career.

Why Sex Matters

I survived after rejecting my colleagues' advice, and in fact, many neuroscientists have come to realize like me that their deeply ingrained assumption that sex does not matter is just plain wrong.

Let us start with animal research. Despite the fact that most neuroscientists still overwhelmingly use only males in their studies, other neuroscientists have generated considerable data demonstrating sex influences on brain function at all levels, including the molecular level[6-8] and ion-channel level.[7] Very often these sex influences are completely unanticipated by investigators. Crucially, animal research clearly demonstrates that mammalian brains in particular are filled with sex influences that cannot be explained by human culture. Thus animal research proves that the human mammalian brain must contain all manner of biologically based sex influences—from small to large—that cannot be explained simply by human culture (even though there are certainly cultural contributions in many cases). Animal research has torpedoed the "it's all human culture" ship that ruled the academic seas since the 1970s when it came to sex differences.

But evidence of sex influences on brain function is not restricted to animal research. Research involving humans has generated equally impressive findings, two of which I highlight here, one regarding human brain structure, the other human brain genetics.[6-8]

One recent landmark study came from investigators from the University of Pennsylvania. They used a form of magnetic resonance imaging (MRI) called diffusion tensor imaging (a way to measure the brain's white matter, or axons by which neurons connect) in a large sample of men and women (428 males and 521 females, ages 8 to 22 years).[9] Across a number of different analytic approaches, they found a striking and consistent result: The brains of women exhibit significantly stronger patterns of interconnectivity across brain regions—including across the hemispheres—than do the brains of men, which conversely exhibit significantly greater average connectivity within local brain regions (what the authors refer to as modularity).

This striking result fits very well with a highly consistent finding across the sex-difference literature: The brains of men tend to be more asymmetrically organized across the hemispheres than are those of women.[10] Importantly, the authors found no age-by-sex interactions despite having plenty of statistical power to find such interactions. This means we cannot explain the sex differences in their results as simply being due to different cultural experiences between males and females.

The Pennsylvania study results are also consistent with diffusion tensor imaging studies by Neda Jahanshad and colleagues, who found greater average interhemispheric connectivity in women compared to men.[11-12] (Impressively, with some analytic approaches, these investigators can accurately classify brain connectivity networks based on sex with 93 percent accuracy.[12]) While we can, and should, debate what these types of anatomical findings will ultimately prove to mean functionally, the evidence leaves little reasonable doubt that male and female brains exhibit, on average, differing patterns of structural interconnectivity, in particular between the hemispheres. In a comprehensive review of human-brain connectivity studies from several years ago, Gaolang Gong and colleagues concluded that "it should be mandatory to take gender into account when designing experiments or interpreting results of brain connectivity/network in health and disease."[13] The data since then confirm this view.

A second important study highlights the fact that sex differences exist down to the genetic level in humans. When David Cribbs and other researchers performed a comprehensive analysis of the patterns of expression in the brain of immune system–related genes in human aging and Alzheimer's disease (AD), they found sex-specific patterns of gene expression in both conditions.[14] In particular, they compared patterns of gene expression in two regions that are critical for higher cognitive function and known to develop AD-type pathology: the hippocampus and a region of the frontal cortex called the superior frontal gyrus. The hippocampus was more prone to immune-type gene reactions in females than in males, while the superior frontal gyrus was more susceptible to immune-type gene reactions in males than in females. Studies such as this prove that the biological mechanisms of brain aging and disease cannot be assumed to be the same in men and women.

The Counter-Reaction

Perhaps not surprisingly, the striking growth in sex-differences research appears to have elicited a counter-reaction from some academic quarters, especially among non-neuroscientists. In some cases this counter-reaction is justified, as when scientists object to gross overstatements about sex differences often made in best-selling books ("neurotrash," as it is sometimes called). But in the main, this counter-reaction appears to reflect a misunderstanding of some key facts of brain biology. Leaving aside the name-callers (such as the psychologist who calls people studying brain-sex differences "neurosexists"[15]), as well as the non-neuroscientists who hypercritically analyze a small fraction of the neuroscience literature while seemingly remaining unaware of the rest,[16] let's focus on key arguments made by "anti–sex difference" authors.

First, anti–sex difference authors argue that there are few (if any) meaningful behavioral differences between men and women. They invariably rely on meta-analyses—studies that analyze patterns across many published studies.[17] Typically, these meta-analyses examine the literature for the size of sex differences (in this case, the size of the difference in average performance between men and women) on isolated factors, such as reading comprehension or the ability to rotate a three-dimensional object in one's mind. And often (though not always), these meta-analyses suggest that, with a few exceptions such as sexual behavior and aggression, only very small (and by implication dismissible) differences exist in the behavior of men and women. But there are at least two problems with these sorts of analyses. First, as Sarah Burnett has illustrated very powerfully,[18] it is simply incorrect to conclude that because an average difference between men and women is quantitatively small, that difference will have few meaningful practical consequences. Second, claiming that there are no reliable sex differences on the basis of analyzing isolated functions is rather like concluding, upon careful examination of the glass, tires, pistons, brakes, and so forth, that there are few meaningful differences between a Volvo and a Corvette.

A more sensible analysis is one that better gauges the full behavioral patterns of men and women. In a fascinating study, Marco Del Giudice and his colleagues did just that.[19] Using a form of statistical analysis called multigroup latent variable modeling, which essentially assessed the size of sex differences by combining numerous isolated factors, they found very large sex differences in behavior with as little as a 10 percent overlap between the distributions of men and women. They powerfully conclude, "The idea that there are only minor differences between the personality profiles of males and females should be rejected as based on inadequate methodology."

Another way to defeat the idea that there are no behavioral sex differences between men and women is to consider stereotypical male and female behaviors. Bobbi Carothers and Harry Reis did just this when they analyzed the size of sex differences in a variety of stereotypically gender-driven behaviors, such as playing golf or video games, watching pornography or talk shows, taking a bath, and talking on the phone.[20] Using this analysis, they report extremely large, bimodal (also called taxonic) sex differences that, as they correctly note, say absolutely nothing about the degree to which those taxonic behaviors result from biological or environmental factors. It may not be fairly assumed that stereotypical behaviors result solely from environmental factors. (Indeed, it has been shown that male and female stereotypical occupational preferences are strikingly consistent across 53 countries, ranging from Pakistan to Norway, under hugely variable cultural conditions.[21]) Carothers and Reis powerfully invalidate the idea that there are no large, group-average sex differences in human behavior outside a few limited domains.[20]

Worse still for the anti–sex difference authors is the fact that a complete, fully-agreed-upon-by-all lack of a sex difference in a particular behavior means absolutely nothing about whether or not sex differences exist in the neural substrates of that behavior. Neuroscientist Geert de Vries most convincingly makes this case, which even his own colleagues occasionally forget.[22] Focusing on a variety of animal models, de Vries shows that sex differences in the mammalian brain often exist to prevent behavioral-level sex differences (by compensating for underlying neural or hormonal differences) rather than to create behavioral-level sex differences. But understanding these compensatory sex differences is every bit as important to properly treating brain dysfunction in men and women as is understanding sex differences that induce behavioral differences.

A second argument that the anti–sex difference authors sometimes make is that there really aren't male and female brains; rather, men and women have a single "intersex" brain. In attempting to support this view, Daphna Joel,[23] who has stated that sex-difference research can make her "blood boil,"[24] correctly points out what neuroscientists have known from animal research since the 1970s or earlier: Both males and females are exposed to both masculinizing and feminizing influences. She also correctly refers to both male and female brains as "mosaics" of such influences—and she is far from the first person to do so.[6] But because most of these influences can vary by degree and circumstances, she concludes, "We all have…an intersex brain (a mosaic of 'male' and 'female' brain characteristics)." The fallacy in her argument lies in the implication that "we all" (men and women) have a *single* mosaic "intersex" brain. What she clearly means by the term *intersex* is "unisex"—there is only one. However, zero evidence supports the view that, through the normal course of development, male and female mammals, including humans, possess brains that have on average the *same* combination of masculine and feminine traits—that they possess a single unisex mosaic brain. Also, the unisex view fails to accommodate a host of facts, such as the remarkable hemisphere differences in X inactivation seen only in female brains, the consequences of incomplete X inactivation (again, only in female brains), direct Y-chromosome-linked effects on brain function in males, or dyslexia's incidence in up to 10 times as many males as females, to name just a few.[25-28] We aren't unisex, and every cell in the brain of every man and every woman knows it.

"But wait," argue the anti–sex difference authors, "the brain is plastic"—that is, molded by experience. One group of authors uses the word *plasticity* in the title of their paper three times to make sure we understand its importance.[29] (As someone who has studied brain plasticity for more than 35 years, I find the implication that it never occurred to me amusing.) By the plasticity argument—also made explicitly by neuroscientist Lise Eliot in her book *Pink Brain Blue Brain*—small sex differences in human brains at birth are increased by culture's influence on the brain's plasticity.[30] Eliot further argues that we can avoid "troublesome gaps" between the behaviors of adult men and women (a curious contradiction, by the way, of the view that there are no behavioral differences between the sexes) by encouraging boys and girls to learn against their inborn tendencies.

It is critical to understand where the fallacies in this argument lie. First, it is false to conclude that because a particular behavior starts small in children and grows, that behavior has little or no biological basis. One has only to think of handedness, walking, and language to see the point. Second, this argument presupposes that human "cultural" influences are somehow formed independent of the existing biological predispositions of the human brain. But third, and most important, is the key fallacy in the plasticity argument: the implication that the brain is *perfectly* plastic. It is not. The brain is plastic only within the limits set by biology.

To understand this critical point, consider handedness. It is indeed possible, thanks to the brain's plasticity, to force a child with a slight tendency to use her left hand to become a right-handed adult. But that does not mean that this practice is a good idea, or that the child is capable of becoming as facile with her right hand as she might have become with her left had she been allowed to develop her natural tendencies unimpeded. The idea that we should use the brain's plasticity to work against inborn masculine or feminine predispositions in the brains of children is as ill conceived as the idea that we should encourage left-handed children to use their right hand.

The presence of biological limits to plasticity—and hence the presence of limits to how much experiences can affect the brain—is perhaps made most clear in elegant studies by J. Richard Udry. In his important but underappreciated paper entitled "Biological Limits of Gender Construction," Udry examines the interaction between two factors—how much a mother encouraged her daughter to behave in "feminine" ways, and how much the daughter had been exposed to masculinizing hormonal influences in the womb—on how "feminine" the daughter behaved when she was older.[31] The figure below illustrates the key findings.

The graph illustrates that, indeed, the more mothers encouraged "femininity" in their daughters, the more feminine the daughters behaved as adults, but only in those daughters exposed to little masculinizing hormone in utero. Crucially, the greater the exposure to masculinizing hormonal effects in utero (the progressively lower lines), the less effective was the mother's encouragement, to the point where encouragement either did not work at all (line with squares) or even tended toward producing the opposite effect on the daughters' behavior (line with diamonds).

All those wishing to understand sex influences on the human brain need to fully grasp the implications of the animal literature, and then think about the Udry data, which captures an incontrovertible fact from brain science: Yes, brains are plastic, but only within the limits set by biology. It is decidedly not the case that environmental experience can turn anything into anything, and equally easily, in the brain. The specious plasticity argument invoked by anti–sex difference authors appears to be just a modern incarnation of the long-debunked "blank slate" view of human brain function, the idea that all people's brains start out as blank slates, thus are equally moldable to become anything through experience.[32]

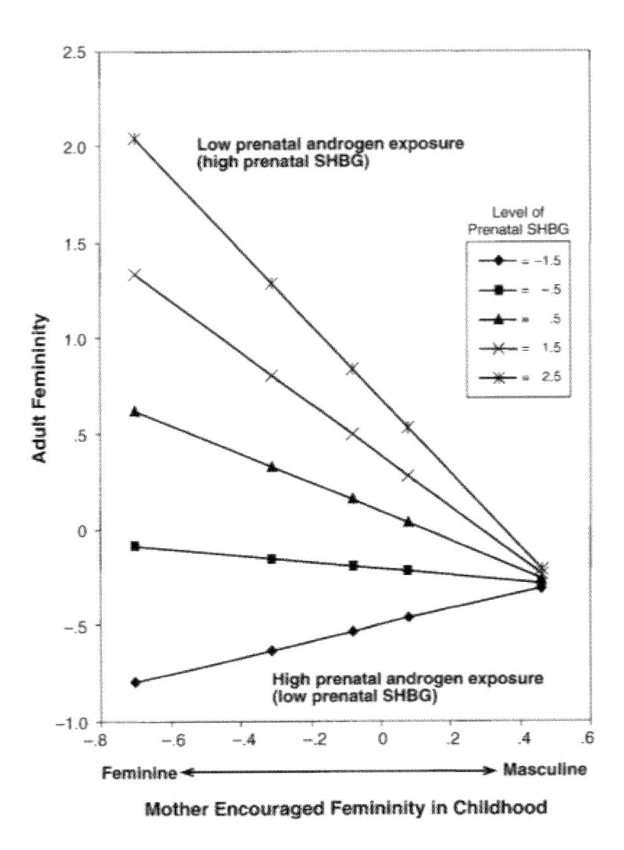

Figure 1. Effect of Childhood Gender Socialization on Adult Gendered Behavior by Level of Prenatal Androgen Exposure. Sex hormones (androgens, estrogens and progestins) operate in both males and females; in the brain, both sexes have receptors for these hormones that are found not just in brain areas associated with reproduction and related behaviors (eg hypothalamus) but in virtually all brain areas, including those involved in higher cognitive functions. These hormones, receptors for which are found not only in the cell nucleus, but also in many other cellular compartments often near membranes, influence many neurobiological events via the genome, epigenome and non-genomic cellular signaling. The process of sexual differentiation involves virtually the whole brain and is a seamless ongoing interaction over the lifecourse between hormones and experiences.

What Darwin Actually Said

We should have expected all along that the brains of men and women are a complex mix of similarities and differences, at least if we believed in evolution as Charles Darwin described it. Darwin did not believe that evolution proceeded by natural selection. In fact, he was completely clear that, in his view, evolution by natural selection alone must fail. He knew that natural selection alone failed to explain far too many phenomena (most famously the male peacock's tail). What Darwin actually said was that evolution proceeded largely through two distinct mechanisms: natural selection and sexual selection. The former acted on the basis of whether an organism survived; the latter acted on whether it made a baby. In his second book, *The Descent of Man, and Selection in Relation to Sex*, Darwin developed this idea (first presented in the original edition of *The Origin of Species*) and made explicit his view that the beneficial effects of sexual selection must at times outweigh the negative effects of natural selection (again, think of the male peacock's tail).

After receiving much criticism for this concept, as he also did for natural selection, Darwin said, "My conviction in the power of sexual selection remains unshaken."[33] Sexual selection is a force that, by definition, often acts male on male or female on female. It is therefore a force that must produce sex differences of many sorts in brain and mind, as Darwin discussed in great detail. Thus, if we believe in evolution as Darwin described it (as a complex mix of natural and sexual selection forces), then we *must* believe that it produced in men and women bodies and brains that are a complex mix of similarities and differences, small to large—exactly what it appears to have produced.

Evolution has produced mammalian brains that are filled with biologically based sex similarities and differences, down to the molecular level. Evolution also has produced in men and women bodies that are filled with similarities and differences, including in the heart, liver, lungs, immune system, and even knees.[34] To insist that somehow—magically—evolution did not produce biologically based sex influences of all sizes and sorts in the human brain, or that these influences somehow—magically—produce little or no appreciable effect on the brain's function (behavior) is tantamount to denying that evolution applies to the human brain.

False Assumptions

At the root of the resistance to sex-influences research, especially regarding the human brain, is a deeply ingrained, implicit, false assumption that if men and women are equal, then men and women must be the same. This is false. The truth is that of course men and women are equal (all human beings are equal), but this does not mean that they are, on average, the same. 2 + 3 = 10 – 5, but these expressions are not the same. And, in fact, if two groups really are different on average in some respect, but they are being treated the same, then they are not being treated equally on average.

Sadly, this is exactly the case in research and medicine today. Women and men are not being treated equally, because by and large women are being treated as if they were the same as men. To make real progress in improving *both* men's and women's health, and to avoid more Ambiens or worse, we need neuroscientist and non-neuroscientist alike to determine whether they too operate on the false assumption that "equal" means "the same." If so, toss that assumption aside. True equality for the sexes demands it.

5

Rich Man, Poor Man

Socioeconomic Adversity and Brain Development

By Kimberly G. Noble, M.D., Ph.D.

Kimberly Noble, M.D., Ph.D., is an assistant professor of pediatrics at Columbia University, where she studies socioeconomic disparities in children's neurocognitive development. Trained as a developmental cognitive neuroscientist and pediatrician, she received her Ph.D. in neuroscience and her medical degree from the University of Pennsylvania. Current research interests include the time course with which socioeconomic disparities in neurocognition emerge in early childhood; the modifiable environmental factors that mediate such disparities; and the brain-behavior relationships that account for these links. Additionally, Noble is involved in several studies of interventions, including a program targeting children's literacy, math, and self-regulation skills, as well as a study of the effects of experimental poverty reduction on cognitive and brain development. Noble was recently named an Association for Psychological Science "Rising Star."

Editor's Note: Here's a disturbing statistic that made headlines this past January: The richest 85 people in the world now hold as much wealth as the poorest half. Keeping in mind the goal of closing the ever-widening gap between the haves and the have-nots, our author examines new research that ties family income level and other factors to brain development. While socioeconomic adversity may not solely determine a child's success later in life, its significant role in helping children develop language, memory, and life skills can no longer be ignored.

THE HUMAN BRAIN HAS BEEN CALLED "the most complex three pounds in the universe."[1] Indeed, this characterization does not seem hyperbolic when we consider that we are born with 100 billion neurons at birth,[2] and that an additional 250,000 to 500,000 new neurons are formed every minute in the first few months of an infant's life.[3] Further, it is not just the number of neurons, but the number of synapses, or connections *between* neurons, that is extraordinary. Synaptic connections become increasingly complex in the first few years of life, and children have 1,000 trillion connections by age three.[4] Early experiences are critical in shaping this process. In the brain, neural circuits that are used repeatedly tend to strengthen, whereas those that are not used are dropped, or pruned. The most vigorous growth and pruning of these connections occur in the first three to four years of life,[5] meaning that the brain is most plastic, or able to make new connections, early in childhood.

For example, studies in the late 1990s revealed that children who learn a second language early (prior to age seven) show neural organization of the second language that is remarkably similar to that of the first language. In contrast, among late learners of a second language, the second language is in effect "stored separately,"[6] which helps to account for poorer pronunciation and grammar in late-second-language learners.

More recently, the effects of early life experience have been applied to the study of the aging brain. As we get older, the function of our nervous system declines. For example, older adults often have difficulty understanding speech as well as their younger counterparts. This difference is particularly salient in environments with substantial background noise, such as cocktail parties. Researchers recently showed that several years of music training early in life can offset this process of auditory decline.[7] Specifically, a moderate amount of musical training in early childhood is associated with faster neural response to speech later in life, decades after the individual last picked up a musical instrument. The research suggests that early experience with music trains the brain to interact more dynamically with sound throughout a person's life.

While early exposure to additional languages or music may lead to beneficial changes in brain development, early adversity can likewise have important but detrimental effects on the brain.

Early Adversity

Children under 18 years of age represent 23 percent of the population, but they comprise 34 percent of all people in poverty. More than one in five children in the United States live in poverty, representing more than 16 million children.[8] Importantly, the definition of poverty is strictly based on family income and the number of adults and children in the home, with no adjustment for geographic location. Thus, based on the most recent federal guidelines, a family with two adults and two children is considered to live below the poverty line if they earn less than $23,550 per year, regardless of where they reside. The guidelines fail to take into account the fact that raising a family in a city like New York or San Francisco is much more expensive than, for instance, raising a family in rural South Dakota.

For that reason, many researchers partly consider the role of multiple socioeconomic factors in addition to income. Socioeconomic status, or SES, incorporates additional objective measures such as parental education and occupation. Sometimes researchers also consider subjective social status, which is an individual's subjective rating of his or her position in the social hierarchy.

Across these different socioeconomic indices, researchers have described marked disparities in a range of important cognitive and achievement measures for children, such as IQ, literacy, achievement tests, and high school graduation rates.[9] Disparities in achievement tend to emerge early and then widen throughout the early elementary school years. For example, by 10 years of age, family SES is an excellent predictor of a child's cognitive abilities: children from higher-SES families tend to perform well above children from lower-SES families—regardless of whether those children had high or low cognitive abilities at age two.[10]

Numerous factors contribute to these SES gaps in cognitive development: nutrition, environmental toxins, home learning environment, exposure to stress, and early schooling. Further, these different pathways are often highly correlated, in that disadvantaged families are more likely to be exposed to multiple risk factors than are advantaged families. As such, researchers find it daunting to tease out the mechanisms behind the SES gap in cognitive development and, in turn, to design effective interventions.

Weighing Other Factors

One way to begin to make sense of the tangled web of inter-correlated mechanisms leading to socioeconomic disparities in cognitive development is to recognize that cognitive development is itself a very broad construct—too broad to be realistically considered as a single outcome. We are therefore better off trying to understand the links between SES and *specific* aspects of cognition.

The field of cognitive neuroscience teaches us that different brain structures and circuits support distinct kinds of cognitive skills. While classic academic milestones such as school graduation can tell us broadly about global effects of socioeconomic disparities on achievement, we know that achievement is actually a complex output of multiple cognitive and socio-emotional systems, such as language, learning and memory, and self-regulation. These distinct cognitive systems are supported by different brain regions and networks. So, while classic measures of academic achievement such as high school graduation must at some level reflect the function of the brain, they are relatively uninformative when it comes to disruptions or disturbances in specific cognitive and neural processes. By taking a cognitive neuroscience approach, we may improve our efforts at providing targeted educational interventions.

This was the approach that my colleagues and I have taken, beginning when I was a graduate student with Martha Farah at the University of Pennsylvania, in a series of studies over the last decade.[11-13] In these studies, we investigated which core cognitive functions were most strongly related to SES. To do so, we recruited children from socioeconomically diverse families and administered a series of cognitive tests designed to tap in to the core systems of language, executive function, visuospatial skills, and memory—systems that are supported by relatively distinct brain circuits. Across studies, children ranged from kindergarten through middle school age. At any one age, of course, some children perform dramatically better than others. We set out to determine the extent to which such disparities in performance could be explained by differences in SES.

The answer, it turned out, was "to a large extent." In general, children from higher-SES homes tended to perform better on most cognitive skills than children from lower-SES homes. However, the disparities were not uniform. Across studies, we found the largest SES disparities in language skills, with more modest differences in children's memory and executive-function abilities. For example, in one study, for each standard deviation increase in SES (operationalized as a composite of parental education, occupation, and income), language improved by more than half a standard deviation, declarative memory skill increased by approximately one-third of a standard deviation, and certain executive-function skills increased by approximately one-quarter of a standard deviation.[11] Similar socioeconomic gradients in these skills have been reported in children in developing countries.[14] More recent work from our lab has suggested that socioeconomic disparities in neurocognitive development emerge very early, with large differences in language and memory development evident before two years of age.[15]

Building on the Findings

Scientists leading other recent investigations of socioeconomic differences in brain structure and function have considered more specific cognitive and neural outcomes, and it has become possible to begin to tease apart the modifiable environmental factors that mediate these links.

Since the greatest socioeconomic disparities are present in language skills, let us turn first to several findings concerning SES disparities in the function and structure of language-supporting regions of the brain. Polish neuroscientist Przemyslaw Tomalski and colleagues recently used electroencephalography (EEG) to examine SES differences in infants' brain function.[16] This technique is widely used by investigators to examine how powerful a child's brain waves are in different locations across the scalp, thus providing some insight into the activity in different brain regions. Their study found that by six months of age, parent occupation and income were already associated with higher-power brain waves in frontal brain regions. Critically, higher-power brain waves in these regions have been associated with better language development at later ages.[17,18] Thus, it is possible that at least one neural signature of growing up in socioeconomic disadvantage may be detectable very early in infancy, well before behavioral measures of discrepancies in cognitive processing may be evident.

In a recent study in our lab, we examined brain volumes in a group of 60 socioeconomically diverse children ranging from 5 to 17 years of age. We found that, as children get older, higher-SES children tend to dedicate relatively more neural real estate to areas of the brain that support language development, in comparison to their lower-SES peers.[19] This suggested to us that something about the experience of growing up in a higher-SES environment likely leads to a greater investment in language-related regions of the brain.

Indeed, this something is almost certainly experience with language itself. It is well established that children from disadvantaged homes tend to hear fewer words—an estimated 30 million fewer words by age three than their higher-SES counterparts, to be precise.[20] Lower-SES mothers are also more likely to speak to their children in a directive rather than conversational manner, and to use less complex speech patterns and fewer gestures.[20-22] It is likely that differences in maternal speech input result in a cascade of effects that are directly relevant for the development of a child's language-supporting cortex during infancy.[21] Much as greater exposure to music may increase an individual's perception of speech years later, greater social engagement with interactive adults may lead children to have improved abilities to perceive and discriminate among speech sounds.[22] Thus, one mechanistic pathway would suggest that socioeconomic disparities result in large differences in quality and quantity of linguistic exposure, which in turn lead to differences in the development of language-supporting brain regions—and, finally, to the often-reported SES disparities in children's language skills.[19]

The Role of the Hippocampus

As described above, SES disparities in children's learning and memory abilities have been also reported, independent of disparities in language. The hippocampus is one brain structure that is critical for memory development, and a number of recent studies have indicated that SES factors are associated with hippocampal size in both children[19,23,24] and adults across the life span.[25,26] Research in both animals and humans suggests that the experiences of stress and neglectful or abusive parents have direct effects on the development of the hippocampus.[27-30] While family stress is certainly not limited to lower SES families, it is often disproportionately felt in more disadvantaged homes. Thus, a second pathway would suggest that SES differences in exposure to stress may operate on the hippocampus to mediate previously described SES disparities in declarative memory processes.[19] Supporting this notion, Joan Luby of Washington University-St. Louis and colleagues recently found that more hostile parenting relationships and family stress accounted for links between income and hippocampal size.

Finally, socioeconomic disadvantage is associated with a decreased ability to regulate cognition[11-13,31] and emotions,[32-34] a critical aspect of school readiness that predicts grades and achievement-test scores from elementary through high school. Recent work from a number of laboratories has demonstrated SES disparities in the neuroanatomic structure and function of the prefrontal and limbic cortical regions that support these skills.[19,35-40] Again, chronic stress has been associated with alterations in the development of this circuitry.[29,41] Thus, a third pathway would suggest that SES differences in exposure to stress may also operate on prefrontal cortex and limbic circuitry, thus mediating previously described SES disparities in self-regulation.[19] For example, New York University developmental psychologist Clancy Blair and colleagues[42] reported that the link between positive parenting behaviors and children's executive function was partially mediated through the stress hormone cortisol, and Nim Tottenham of Columbia University recently showed that early adversity in the form of maternal deprivation leads to premature adult-like connections between prefrontal and limbic regions.[41]

Thus, mounting evidence suggests that socioeconomic factors—parental education or family income—may lead to differences in the home-language environment or exposure to family stress, which in turn have cascading effects on the development of brain systems that support critical neurocognitive functions such as language, memory, and self-regulation. And yet we still do not know the level at which it is most efficacious to intervene.

Closing the Gap

Are our efforts best directed at improving disadvantaged children's educational experiences, with a focus on language, memory, and self-regulation skills? School-based interventions are certainly the most prevalent form of early-childhood intervention, and many, such as the Chicago School Readiness Project[43] and Boston's pre-K program,[44] have shown promising gains in both preacademic and self-regulatory skills for disadvantaged children. And yet, while these programs can be effective, they are unlikely to be sufficient: Given the size of the SES gap by the time children enter school, preschool interventions alone are unlikely to bridge the gap fully.[45] Some small, intensive early childhood programs such as Perry Preschool or Abecedarian have been shown to result in substantial long-term benefits on cognitive development and achievement, and even physical health as children enter adulthood.[46,47] However, the pragmatics of scaling up such programs to the larger population while maintaining high quality is a frequently cited concern.

Young children spend the vast majority of their time with their parents and other caretakers, and so perhaps we should we be focusing on parents' behaviors. Highly educated parents invest far more time playing with, talking to, and teaching their children, and parenting style has been cited as the single most important factor in explaining the SES gap in cognitive development.[48] And so, perhaps targeting parenting would be the most effective avenue of intervention. Small-scale interventional efforts to teach disadvantaged parents about the benefits of speaking early, often, and richly to their children are producing promising results.[49] However, while some larger-scale parenting interventions such as the Nurse-Family Partnership program have led to moderate improvements in children's cognitive and behavioral outcomes, [50] many have a mixed record of success,[51] often due to difficulties with attrition and low participation. Overcoming obstacles to scaling up such interventions will require researchers and policy makers to carefully consider parental motivations and beliefs.[45]

Finally, let us consider interventions that operate at the most distal level—that of SES itself. Correlational evidence suggests that, for disadvantaged families, a $4,000 increase in family earnings in the first two years of a child's life leads to remarkable differences in that child's adult circumstances, including a 19 percent increase in adult earnings, a marked increase in hours spent in the workforce, and even some evidence of improved physical health in adulthood.[52,53] While family income alone is unlikely to be the most important factor in setting young children along an achievement trajectory, it may well be the most malleable factor from a policy perspective. Thus, based on the evidence described above, many leading social scientists and neuroscientists believe that policies that reduce family poverty would have meaningful effects on early caregiving and reductions in family stress, ultimately improving children's brain functioning and promoting the cognitive and socio-emotional development that is so critical for children to succeed and to lead healthy, productive lives.

6

One of a Kind

The Neurobiology of Individuality

By Richard J. Davidson, Ph.D.

Richard J. Davidson, Ph.D., is the William James and Vilas Research Professor of Psychology and Psychiatry, director of the Waisman Laboratory for Brain Imaging and Behavior, and founder of the Center for Investigating Healthy Minds, Waisman Center, at the University of Wisconsin-Madison. Named one of the 100 most influential people in the world by *Time* magazine in 2006, Davidson is co-author (with Sharon Begley) of the *New York Times* best seller, *The Emotional Life of Your Brain* (Penguin, 2012) and founding co-editor of the new American Psychological Association journal EMOTION. He is the recipient of a National Institute of Mental (NIMH)Health Research Scientist Award, a MERIT Award from NIMH, an Established Investigator Award from the National Alliance for Research in Schizophrenia and Affective Disorders (NARSAD), a Distinguished Investigator Award from NARSAD, the William James Fellow Award from the American Psychological Society, and the Hilldale Award from the University of Wisconsin-Madison. He received his Ph.D. from Harvard University in psychology and has been at Wisconsin since 1984.

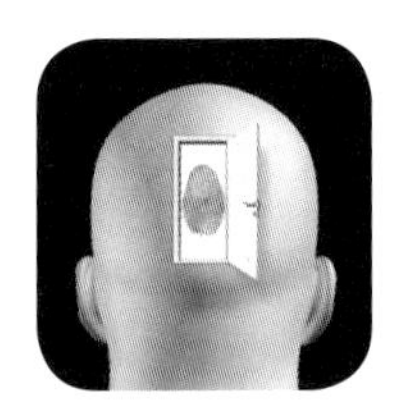

Editor's Note: What processes in the brain are responsible for individuality? Mounting imaging evidence suggests that brain circuits involved in our emotional responses are highly plastic and change with experience, affecting our temperament. Moreover, new research suggests that psychological interventions can further harness brain plasticity to promote positive behavioral changes that increase resilience and well-being.

WHEN WE REFLECT ON THE PEOPLE WE KNOW—family members and individuals in our immediate social or occupational groups—we are often struck by the diversity in personality, temperament, and responses to everyday challenges. Individual variation is perhaps most salient in the realm of emotion, given that our emotions primarily determine how we respond to life's slings and arrows and where we stand on the continuum of psychopathology and resilience. Increasing evidence also suggests that this variation in the emotional response of individuals to common challenges in everyday life is related to peripheral biology—biology below the neck—that may be consequential for physical health.[1]

When attempting to identify the causes of variations in individuality, we must distinguish between distal and proximal causes. Proximal causes typically feature specific brain mechanisms, neural circuits, and molecular processes that underlie the behavioral phenotypes (composites of observable traits) that we measure. Distal causes might include early learning or genetic factors that modulate neural circuits and specify starting conditions or baseline levels of activation in the proximal neural networks that directly control the behavioral phenotypes.

Studies of intra-pair variation in monozygotic (MZ; identical) twins afford an unusual opportunity to characterize variation that is due entirely to non-genetic causes since the two co-twins are genetically identical. A novel study involving 80 MZ twins examined epigenetic variation as a function of age.[2] The researchers found that early in life the co-twins were virtually indistinguishable epigenetically. However, with age, increasingly pronounced epigenetic differences emerged. The authors noted that the fact that epigenetic markers were more distinct in MZ twins who were older, had different lifestyles, and spent less of their lives together underscores a significant role for environmental factors in shaping a common genotype into a different phenotype.

Other prominent behavioral phenotypes have also received extensive scientific attention. Behavioral inhibition, sometimes called anxious temperament, is a phenotype that has been studied in rodents, nonhuman primates, and humans, and is typically associated with high levels of freezing (inhibition of behavior, with the organism remaining relatively fixed in position, not moving and exhibiting high levels of vigilance), decreased vocalizations, and an increase in the stress hormone cortisol release (or corticosterone in rodents).[2] Behavioral inhibition early in life is a predictor of later psychopathology and of increased activation in limbic circuits that play a role in adult anxiety.[3] A related phenotype is associated with rapid recovery from a negative event. Individuals with a slow recovery rate are considered vulnerable, while those with a rapid recovery rate are thought to be resilient.[4] Finally, a third phenotype involves the extent to which a person maintains a positive affect, a characteristic that is central to understanding the underlying affective dynamics of depression.[5]

Nonhumans and Individuality

Studies in nonhuman species have been extremely important in helping scientists to identify some of the causes and consequences of individual differences in aspects of emotional processing. In a unique recent study, Julia Freund and colleagues studied 40 inbred genetically identical mice that lived in a highly enriched environment for three months beginning at four weeks of age.[6] With the goal of studying exploratory behavior, the team computed a measure that they called roaming entropy. High roaming entropy relates to exploring a wide range in a more complex environment. Low roaming entropy is associated with returning to the same location on repeated occasions.

At the end of the experiment, the team assessed hippocampal neurogenesis by counting proliferating precursor cells that they had labeled with bromodeoxyuridine (BUdR) three weeks earlier. Mice raised in this highly enriched environment displayed substantially increased neurogenesis compared with a control group. Most important, mice showing higher levels of roaming entropy also showed greater levels of neurogenesis. The fact that genetically identical mice were used in this study and that their behavior was quite similar across animals at the start of the experiment suggest that experience-dependent changes can induce profound alterations in brain function and structure. The team did not examine why some mice displayed increased roaming entropy over time and others did not, but the study underscores the potency of experience-dependent plasticity in producing individual differences in emotion-related behavior.

In a series of studies of rats bred to show either high levels of locomotor activity to novelty or low levels of locomotor activity in response to the same novel environment (an animal index of exploration/anxiety), Huda Akil and her colleagues established that low responders exhibit significantly fewer ultrasonic vocalizations, which are markers of positive affect; upon repeated exposure to environmental complexity, the low responders change their phenotype and show increases in ultrasonic vocalizations.[7] Akil's study also demonstrated microRNA differences in limbic brain regions between the high and low responder groups.[8] Such microRNAs are known to be potent regulators of gene expression. These findings suggest ways in which these two phenotypes might be associated with differences in gene expression in specific limbic regions.

In a series of collaborative studies with Ned Kalin, M.D., at the University of Wisconsin, we developed a nonhuman primate model of behavioral inhibition. We found that if rhesus monkeys are exposed to the profile of an unfamiliar human, they exhibit freezing. Substantial individual differences exist in the duration of freezing. Rhesus monkeys and human toddlers share two additional features of this phenotype: vocalizations and the steroid hormone cortisol. High freezers show fewer vocalizations, while the stress hormone cortisol demonstrates freezing's positive effects. We created a composite score of behavioral inhibition by standardizing these three metrics and averaging them. The distribution of the composite scores in a large sample is approximately normal.

We also found that individual differences in this composite are reasonably stable over time in rhesus monkeys.[9] By injecting a radiolabeled glucose tracer while exposing the animal to the natural stress of the profile of a stranger, and then placing the monkey in a special nonhuman primate positron-emission tomography (PET) scanner after approximately 30 minutes, we were able to measure brain function during exposure to the stressor, since the images we obtained reflected the integrated activity from the previous 30-minute period. Using this method, we found that metabolic rate in both the amygdala and the anterior hippocampus predicted the extent of anxious temperament and behavioral inhibition.[3]

More recently, in a very large sample, we replicated these basic effects and established that metabolic rate only in the anterior hippocampus (not in the amygdala) was significantly heritable, a finding that was initially surprising.[4] In retrospect, however, we reasoned that it is likely that the amygdala is highly plastic and responsible for various aspects of emotional learning, therefore showing a weaker heritability signal than the hippocampus.[11]

In other recent work, scientists found that animals with stronger resting-state connectivity (measured with resting-state functional MRI) between several regions of the prefrontal cortex, including medial and dorsolateral prefrontal sectors and the amygdala, show lower levels of glucose metabolism in the amygdala. In turn, higher levels of dorsolateral-prefrontal-amygdala connectivity are associated with decreased anxious temperament. This relationship is mediated through decreased glucose metabolism in the amygdala.[12] The collective findings in studies in monkeys indicate that anxious temperament is brought about by a distributed neural network that includes the amygdala and the prefrontal cortex. Several regions of the prefrontal cortex play an important role in emotion regulation while also modulating activation in the amygdala. It is also likely that these regions modulate the time course of amygdala response.

I believe that future researchers would profit by examining experience-dependent amygdala plasticity in humans. Our studies in nonhuman primates suggest that individual differences in amygdala function and the associated circuitry directly interconnected with the amygdala, including the prefrontal cortex, the bed nucleus of the stria terminalis, the anterior hippocampus, and the periaqueductal gray play a key role in determining individual differences in anxious temperament. All of these areas likely play an important role in governing individual differences in both reactivity to and recovery from negative events.

Our studies in humans, which have begun to parse the temporal course of emotional response, are predicated on the intuition that resilience is, at least in part, associated with rapid recovery following adversity, while vulnerability is associated with the opposite—a difficulty in recovering from negative events. Temporal dynamics are also important in the realm of positive affect. Individuals who can savor and sustain positive affect may show higher levels of well-being than those who cannot.

A Matter of Timing

One central characteristic of resilience may be more rapid recovery following negative events. The recovery may occur in any of several different systems that show stress-related reactivity. Two individuals may respond equally but at different rates. This is illustrated below in the hypothetical curves in figure 1. We have measured the time course of response to emotional stimuli with both peripheral physiological measures and more direct measures of brain function.

In a recent study, we showed that people who reported higher levels of well-being, particularly regarding "purpose in life," faster and more complete recovery following negative stimuli.[5] In another study, we found that those who recover more quickly also report higher levels of conscientiousness.[6] In each of these studies, we controlled for how much reactivity the person showed to the emotional stimulus and thus we were able to obtain a "pure" measure of recovery. In yet another recent study, we used functional magnetic resonance imaging (fMRI) to identify neural correlates of recovery and found that variations in the time course of recovery of activation in the amygdala was a predictor of individual differences in neuroticism, one of the best-studied traits reflecting negative emotion.[7] The most important finding in our study was that variations in reactivity—the initial responses to the negative stimuli—did not predict neuroticism. Collectively these

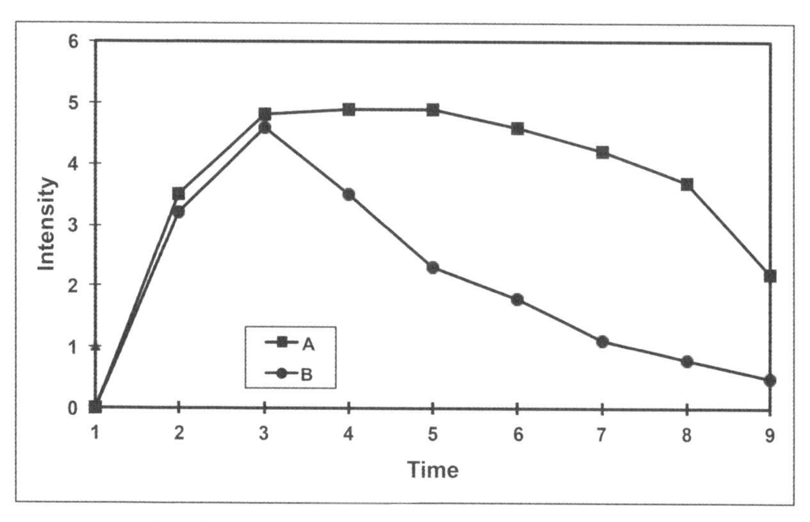

Figure 1: Hypothetical curves showing reactivity and recovery for two individuals, with each person showing comparable amplitude of response but with person B recovering more quickly than person A.

findings underscore the importance of individual differences in recovery from negative events as a key constituent of variations in emotional style and suggest that the time course of amygdala response likely plays an important role in moderating these effects.

In another series of studies, we have examined variations in positive affect, with specific interest in the extent to which individuals persist in positive emotional response. Some have called this persistence "savoring" or "sustainment."[8] We found that patients with major depression show normal levels of initial activation in ventral striatal regions but fail to sustain this activation over time. The less the depressed patients were able to sustain ventral striatal activation over time, the lower were their reported levels of positive affect.[5] Patients who show increases in sustained activation of the ventral striatum due after two months of antidepressant treatment also show increases in treatment-related gains in positive affect.[16] And in a community sample, individuals with greater sustained activation of the ventral striatum report higher levels of psychological well-being and have lower basal levels of cortisol.[16] This latter finding indicates an important association between patterns of central nervous system activity that are associated with emotional styles and peripheral biological processes that are implicated in physical health.

An abundance of evidence suggests that high levels of well-being are associated with better physical health.[18] The mechanisms by which this association arises are unknown, but its existence is now well documented. These findings raise the possibility that interventions that are designed both to promote well-being and to influence the central circuitry of emotional style may also have peripheral biological benefits on physical health.

Our findings also indicate that one important constituent of individual differences in positive affect and well-being is the ability to sustain positive affect following a positive incentive. Depressed patients show activation in circuits important for positive affect following the initial presentations of positive stimuli, but they fail to sustain this activation. Those who can sustain such activation report higher levels of positive affect and higher levels of well-being. In addition, in a community sample, such individuals have lower serum levels of cortisol. The temporal dynamics of emotional response play an important proximal role in modulating individual differences in emotional response. We do not yet know the distal causes of these individual differences, though some combination of genetic and environmental factors clearly plays an important role.

Plasticity and Individuality

The brain circuits that underlie individual differences in emotional response and emotion regulation are highly plastic and can be altered in an experience-dependent fashion—that is, they change in response to interventions.[19] Scientists have found that several nonpharmacological interventions that are designed to reduce anxiety and depression and to promote well-being change the central circuitry of emotion both functionally and structurally.[19] For example, Britta Hölzel and her colleagues found that among participants going through mindfulness-based stress reduction, the greater the decrease in perceived stress, the greater the reduction in amygdala volume over the course of the eight-week intervention.[20]

A recent study randomized patients with Parkinson's disease (PD) either to an eight-week mindfulness meditation intervention or to usual care. Structural MRI was obtained before and after the eight weeks. Mindfulness meditation was associated with increases in gray-matter volume in the caudate and other related regions implicated in PD compared with the control condition.[9] Thus, even among patients with a frank neurological disorder, benefits may result from a nonpharmacological intervention that targets some of the key circuitry of the regulation of attention and emotion. Unfortunately, measures of emotional style were not obtained in this study, and so I prefer not to comment on the relationship between structural brain changes and emotional aspects of individuality.

Our lab recently published a study that evaluated the impact of a short-term intervention designed to cultivate compassion and the associated positive emotions linked to compassion.[10] Participants were randomly assigned to either a compassion-training intervention or a cognitive-reappraisal training intervention that was structurally matched to the compassion training. Functional MRI was obtained before and after the two-week interventions. In addition, at the conclusion of the interventions, both groups were administered an economic decision-making task to assess individual differences in altruistic behavior. We found that after two weeks of compassion training, those assigned to the compassion-training group showed significantly more altruistic behavior compared with the group assigned to the cognitive training. In addition, we found systematic alterations in brain function that predicted the increase in altruistic behavior. One key change in brain function was an increase in connectivity between the dorsolateral prefrontal cortex and the nucleus accumbens, which showed enhanced connectivity for the compassion group compared with the cognitive-reappraisal-training group. This enhanced connectivity in the compassion group predicted increased altruistic behavior. This prefrontal-striatal network is the same as the one implicated as deficient in depression, and, again indicates that the circuits underlying emotional styles are at least somewhat plastic and can be altered through training.

Individuality, particularly in the realm of emotional responding, provides color to our everyday life and infuses our interpersonal relationships with meaning. The fact that the brain networks that underlie such individuality exhibit plasticity is not surprising, for we all recognize that early adversity can result in long-term deleterious effects for a person. However, the very plasticity that can cause pathology is also the source of potential positive change. We can harness the potential of plasticity to shape the brain in more intentional ways to cultivate healthy habits of mind that can confer resilience. The prospects of having this perspective be widely recognized and adopted is personally very significant to me, for I believe that if we all took more responsibility for our minds and brains in these ways by intentionally cultivating healthy habits of mind, we can exercise the brain in ways that are similar to exercising the body and potentially promote positive behavioral changes that might increase resilience and well-being in a large fraction of the population.

7

The Age Gauge

Older Fathers Having Children

By Brian M. D'Onofrio, Ph.D., and Paul Lichtenstein, Ph.D.

Brian M. D'Onofrio, Ph.D., is a professor and the director of clinical training in the Department of Psychological and Brain Sciences at Indiana University. He received his Ph.D. in clinical psychology from the University of Virginia in 2005. His research explores the causes and treatments of psychiatric problems using advanced epidemiological research designs and longitudinal analyses. He has received funding from the National Institutes of Health, the Swedish Research Council, Indiana University, the Indiana Clinical and Translational Sciences Institute, and the Brain & Behavior Research Foundation. D'Onofrio has also received several awards for his research, including awards from the Association for Psychological Science, the Behavior Genetics Association, the Brain and Behavior Research Foundation, Indiana University, and the Society for Research in Child Development.

Paul Lichtenstein, Ph.D., is a professor in genetic epidemiology and also head of the Department of Medical Epidemiology and Biostatistics at the Karolinska Institutet in Sweden. He is a world-renowned twin and family researcher and an expert on using the unique Swedish population-based registers to answer questions regarding the causes and consequences of mental-health problems. He has won many awards, including the James Shields Award for Lifetime Contributions to Twin Research in 2011. Lichtenstein is funded by several Swedish agencies, as well as international funding sources such as the National Institutes of Health and the European Union.

Editor's Note: In recent years, scientists have debated the existence of a link between a father's age and his child's vulnerability to psychiatric problems. Our authors led a research team that produced a paper that analyzed data on all individuals born in Sweden from 1973 through 2001. Both the authors' study and another study raise as many questions as they answer, but they suggest that children born to middle-aged men are more likely than their older siblings to develop a range of mental difficulties, including bipolar disorder, autism, and schizophrenia.

"WHEN SHOULD I HAVE CHILDREN?" "Am I too old to have a child?" "How old is too old?" "Is my daughter destined to have problems because I had her when I was older?" These are just some of the questions that many people, including colleagues, friends, and anonymous individuals (via email), posed to both of us after the news media covered the results of two large-scale studies on the association between paternal age at childbearing and mental-health problems of offspring. One study, which was conducted by John McGrath and colleagues, was based on data from Denmark.[1] The second study, based on data from Sweden, was conducted by the two of us and our colleagues.[2] How did we answer these questions from the public? What does the science say—and not say—about the topic?

At the outset, let us first acknowledge that these two studies address very sensitive issues. Decisions about whether to have children and when to have children are incredibly private. In addition, the prospect that one's children will have both mental-health and physical problems is a major concern for parents, and caring for offspring with such problems can cause considerable pain and suffering.

Overview of the Studies

The two recent studies had many similarities. First, both analyzed comparable data sets to explore the associations with parental age at childbearing via epidemiologic methods that use large, representative samples to examine the distribution and determinants of health. The data sets in both studies were based on extensive population records kept by the governments of Scandinavian countries, including Denmark and Sweden. For example, these countries have extensive records of all inpatient and outpatient visits to hospitals. Scandinavian governments also are able to use personal identification numbers to link several databases. As a result, separate databases can be combined such that they contain information across different generations and domains (e.g., psychiatric diagnoses and age at childbearing). To facilitate the analysis of these population data sets while protecting personal information, government agencies in both countries alter the personal identifiers after merging the data sets, a practice that has led to extensive psychiatric epidemiology research.[3,4] Unfortunately for research purposes, no such data sets are available in the United States.

Second, both studies took advantage of the size and scope of the national data sets. Access to these large data sets enabled both research groups to explore a fuller range of parental age at childbearing because the data sets included high numbers of individuals at the extreme ends of childbearing (e.g., older than 45 years). The large data sets had sufficient numbers of exposed cases (e.g., offspring diagnosed with autism whose parents were older than 45 years at childbirth), enabling both groups to predict relatively rare outcomes in the offspring (e.g., schizophrenia and autism). Previous research teams, which relied on smaller data sets, could not explore advancing paternal age at childbearing with as much precision.

Third, both studies presented population and adjusted estimates of the magnitude of the association between paternal age at childbearing and disorders of offspring. Notably, the two estimates address different research questions. The population estimates respond to this question: How common is it that an offspring born to an older father has psychiatric problems? Both research teams sought to answer this question by comparing the rates of several disorders among offspring born to fathers of different ages in the entire population. The adjusted estimates, on the other hand, respond to this question: When it comes to psychiatric problems in offspring, how much risk is *specifically* due to advancing paternal age?

The studies addressed the latter question by using statistical techniques and other approaches to estimate the magnitude of the association between paternal age at childbearing and offspring disorders while trying to hold constant the highest possible number of factors that differ among fathers who have children at different ages (e.g., parents' socioeconomic status). Researchers must use various techniques to study advancing paternal age at childbearing because men who have children when they are younger differ, sometimes greatly, from men who delay childbearing. Plus, it is impossible to conduct a randomized experiment in humans—you can't randomly assign men to have children at different ages.

It is important to stress that both research questions (based on population and adjusted estimates) are valid and might be of interest for public-health policy. Both questions can also have important implications for subsequent basic research because the analysis of large population data sets can provide critical information about mechanisms at multiple levels of analysis that researchers should explore in the future.[5–9]

Previous epidemiological research on advancing paternal age had suggested that advancing paternal age is associated with increased risk for psychiatric problems in the offspring, but the findings were inconsistent.[10–16] Furthermore, many researchers suggested that any association between advancing paternal age and offspring psychiatric problems was not due to the specific consequences of delaying childbearing; rather, differences associated with advancing paternal age, such as personality and psychiatric problems in the fathers[17–19] and birth-order effects in the offspring,[11] could better explain any associations. Given these concerns, the two recent studies based on the large, population-based registries were intended to provide more understanding of the associations between paternal age at childbearing and offspring psychiatric problems.

The Danish Study

The study by McGrath and colleagues included all individuals born in Denmark from 1955 through 2006, a sample of almost 3 million people. The study indexed psychiatric problems based on inpatient hospitalizations and outpatient visits to psychiatric departments. The study presented population estimates for several psychiatric disorders. For instance, researchers found that compared to offspring born to fathers 25 to 29 years old (the reference group in the study), offspring born to fathers over the age of 45 were 1.4 times more likely to have schizophrenia and 1.7 times more likely to have autism. However, the offspring of older fathers were less likely to have hyperkinetic disorders, which is similar to the diagnosis of attention-deficit/hyperactivity disorder (ADHD): Offspring born to men 30 to 34 years old (21 percent less likely), 35 to 39 years old (26 percent less likely), 40 to 44 years old (21 percent less likely), and older than 45 years old (5 percent less likely) had lower rates of these disorders than offspring in the reference group.

In order to provide adjusted estimates that assessed the magnitude of the association between paternal age at childbearing and the offspring disorders that was independent of other factors, the researchers ran a series of analyses that statistically controlled for multiple factors that are correlated with advancing paternal age. For example, the researchers statistically controlled for maternal age at childbearing because men who have children when they are older are more likely to have children with women who are older. In those models, advancing paternal age was still associated with the disorders of offspring, and the magnitudes of the associations were as large or larger than the population estimates. The researchers also accounted for urbanization at place of birth and family history of mental illness. The same pattern or results emerged. For example, offspring born to men over the age of 45 were 1.4 times more likely to have schizophrenia and 1.7 times more likely to have autism. Notably, offspring born to fathers older than 45 were 1.2 times more likely to have a hyperkinetic disorder when statistically controlling for the other factors, despite the fact that in the population, offspring born to men above age 45 were less likely to have the disorder.

The results from the Danish study indicate that in the population offspring of older fathers, some psychiatric problems, such as schizophrenia and autism, are more likely, and that hyperkinetic disorders are less likely. In trying to examine the pattern of associations when accounting for correlated factors, such as maternal age at childbearing and family history of psychiatric problems (the adjusted estimates), the researchers found that advancing paternal age at childbearing was still associated with psychiatric problems. In other words, these measured factors do not explain the associations between advancing paternal age and offspring psychiatric problems. The differences between the population and the adjusted estimates also indicated that researchers must take into account other differences among men who have children at different ages.

The Swedish Study

Our study—conducted separately from the Danish study—included all individuals born in Sweden from 1973 through 2001, a sample of just over 2.6 million people. We similarly explored psychiatric problems, indexed by inpatient hospitalizations and outpatient visits, but our study also included information regarding criminal convictions, low academic achievement based on school grades at age 15, and dropping out of school early. The first analysis provided population estimates. We found that the population of offspring born to older fathers (above 45) had higher rates of schizophrenia (1.6 times more likely) and autism (1.4 times more likely) compared to offspring born to men 20 to 24 years old (the reference group in the study). For many of the other outcomes, however, offspring born to older fathers had fewer problems than offspring in our reference group. For instance, offspring born to men over the age of 45 were 43 percent less likely to have a diagnosis of ADHD.

Our team also used statistical techniques to obtain adjusted estimates that were independent of factors that could correlate with advancing paternal age, but we controlled for a more extensive list of measured variables than the Danish study. We accounted for maternal age at childbearing, as well as paternal and maternal nationality, highest level of education, lifetime history of serious psychiatric conditions, and lifetime history of criminality. Similar to the results in the Danish study, the associations between advancing paternal age and the outcomes in the adjusted group were as large or larger than the overall population estimates. The results indicated that the associations could not be explained by these factors; rather, when accounting for these measured variables, the associations were in some cases larger than the population estimates.

In addition to using statistical techniques to account for differences among fathers who had children at different ages, we also used several advanced research designs to help account for factors that could bias our estimates. In particular, we conducted a sibling-comparison analysis, which estimated the association between paternal age at childbearing and the outcomes while comparing offspring born to the same father. We explored the rates of problems in offspring born when the father was younger compared to their siblings born when the father was older. This design accounted for (or held constant) all traits that made siblings similar, including unmeasured environmental and genetic factors, thus arguably providing a more precise estimate of the specific association with advancing paternal age.[9,20,21]

When we compared siblings, the magnitude of the associations with each outcome was as large as, and sometimes quite larger than, the estimates in the other analyses. For instance, offspring born to fathers over the age of 45 were 2.1 times more likely to have schizophrenia and 3.4 times more likely to have autism than their siblings who were born when their father was younger. Furthermore, whereas the population estimates found advancing paternal age to be correlated with lower incidence of ADHD, the sibling comparisons indicated that advancing paternal age was more strongly associated with the disorder (for example, offspring of fathers over the age of 45 were 13 times more likely to have the disorder than their siblings born when the father was 20 to 24 years old). The sibling-comparison approach has many advantages, but the design also has several limitations (for example, do the results apply or generalize to other populations?).[9] To help address several concerns, we conducted numerous additional analyses, including the comparison of cousins and firstborn cousins, and found comparable results: advancing paternal age was associated with more psychiatric problems, and the magnitude was stronger than the population estimates.

The population estimates from our Swedish study were generally consistent with the findings from the Danish study: Advancing paternal age was associated with greater risk for some disorders, such as schizophrenia and autism, but less risk for others, including ADHD. The adjusted estimates (when controlling for measured traits of both parents and when comparing siblings and cousins born at different ages) suggested that advancing paternal age was even more strongly associated with psychiatric problems than previous estimates indicated.

Understanding the Underlying Processes

What could explain the finding that advancing paternal age at childbearing is associated with more offspring psychiatric problems? The working hypothesis that guided both studies was that genetic mutations during the production of sperm, referred to as de novo mutations, increase as men get older and have a causal influence on offspring psychiatric problems. Unlike in women, who are born with all of their eggs, in men sperm continue to replicate throughout their lifetime. In fact, sperm cells undergo 20 to 30 divisions per year—approximately 600 divisions by the age of 40.[22] Each cell division brings the possibility of new mutations. Recent studies suggest that there are approximately two new mutations each year, and there is an exponential increase where mutations double every 16.5 years.[23] A growing number of molecular genetic studies have found that these de novo mutations play a large role in human diseases, including psychiatric problems.[24] For instance, several studies have indicated that de novo mutations are associated with autism, suggesting a mediating biological pathway that could explain the association between advancing paternal age and the disorder.[25–27] The two recent epidemiological studies we reviewed above, which found independent associations between advancing paternal age and offspring psychiatric problems, are consistent with the role of de novo mutations.

But why would the adjusted estimates in the studies be larger—sometimes quite larger—than the population estimates? To understand the differences in magnitude of these two types of estimates, it may be important to understand the genetic, psychological, educational, social, and financial context associated with advancing paternal age because many of these factors predict fewer psychiatric problems in the offspring. Twin, adoption, and family-based studies have clearly shown that genetic factors influence age at first childbearing. There are no genes "for" early or late childbearing; rather, genetic factors influence personality traits, psychiatric problems, and other characteristics that in turn influence age at childbearing.[28] For instance, a recent family study found that women who have ADHD, their siblings who did not have the disorder, and men who have children with women who have ADHD are all more likely to have children as teenagers.[29] Offspring born to older parents, therefore, have lower genetic risk for ADHD on average than offspring born to parents who were teenagers at childbearing. Furthermore (as we also showed in our study), advancing paternal age at childbearing is also correlated with higher levels of parental education and higher family income, both of which lead to increased social and cultural capital.[30] In sum, delaying childbearing is correlated with a host of factors that also predict fewer psychiatric problems in offspring.

Thus, one conceivable explanation for the discrepancies between the population and adjusted estimates stems from the possibility that the population estimates could be an amalgam of the deleterious effects associated with de novo mutations and the protective factors (e.g., genetic inheritance, personality traits, and social/cultural capital) associated with delaying childbearing. The models that adjusted for measured traits and compared siblings and cousins, therefore, may have held constant many of these protective factors. As a result, the analyses may have provided a clearer estimate of the specific influence of delaying childbearing.

Implications of the Research

The Danish and Swedish studies are examples of translational epidemiology, which can help guide subsequent basic research.[5–9] The provocative findings regarding advancing paternal age at childbearing suggest that more research needs to be conducted on de novo mutations, especially given recent technological advances that enable researchers to better measure and characterize genetic mutations.[24] It is important to note, however, that there could be other mechanisms through which advancing paternal age at childbearing comes to be associated with offspring psychiatric problems. Therefore, research into other biological and social factors is needed to better clarify the processes that account for the findings in these studies. Furthermore, the two studies also highlight the need to understand the complex and multifaceted factors that are associated with delaying childbearing.[31]

So, finally, what do these studies mean for people who are making decisions about childbearing? How did we answer the questions we received? As you might imagine, we were reluctant to provide concrete advice. We were not evading the questions. Rather, we do not think science can provide a definitive answer to these questions. (Plus, it is unethical to give medical advice to strangers via email.) But here is how we responded to the questions:

First, these are only two studies. Our study in particular, which was one of the first studies to use sibling and cousin comparisons to study advancing paternal age, needs to be replicated. We think it was a good study (we know we are biased), but there is never one definitive study, especially in this area of research, because each and every study has limitations. Again, we cannot conduct randomized controlled studies of paternal age at childbearing. As a result, researchers must obtain consistent findings using many different designs and samples before they can make strong causal inferences (e.g., advancing paternal age causes offspring psychiatric problems), especially regarding implications for family policy.[32]

Second, researchers need to conduct more studies on the topic before any professional group can make explicit recommendations. In particular, researchers must expressly examine if (and then how) physicians and couples could incorporate information on paternal age at childbearing into their decision-making process. This examination requires studies using predictive models.

Third, both studies provide evidence that the overwhelming majority of offspring born to older dads will not have a major psychiatric or related problem. Both studies found that the risk that a child will have psychiatric problems was correlated with advancing father's age, but most of the outcomes were quite rare. For instance, in our study less than 3 percent of the offspring had psychiatric problems. Therefore, advancing paternal age does not mean that any particular child will definitely develop problems, nor does it mean that a psychiatric problem in an offspring born to an older father was actually caused by the father's age at childbearing.

Fourth, our study and others also have indicated that there can be advantages to delaying childbearing. Some factors that are correlated with delaying childbearing (e.g., attaining a higher level of education or gaining financial security) predict better outcomes in children. These two studies, therefore, add to a growing body of research suggesting that families, doctors, and society as a whole must consider both the potential advantages and the potential disadvantages of delaying childbearing.

Fifth, when trying to weigh the possible risks and benefits of delaying childbearing, there is no set age at which advancing paternal age suddenly becomes problematic. Our study's reports of increased risk for offspring born to men over the age of 45, for example, were just one illustration. Both studies found increasing risk for many of the disorders as paternal age increased, referred to as a dose-response relationship (i.e., offspring born to men 40 to 44 years old also had higher rates of the disorders compared to the reference groups). In fact, each research paper provides graphical representations of magnitude of the associations between paternal age at childbearing and the outcomes across the entire range of parental age.

Sixth, individuals and couples concerned about the consequences of delaying childbearing should consult with their physician or a specialist in their area. For instance, concerned individuals and couples can meet with genetic counselors, who can provide more detailed and personal information regarding the risks associated with delaying childbearing.

Finally, and most important, there are many personal circumstances and values that go into making the decision of when to have a child or children. Yes, we think that research can help inform personal decision-making. But no study, set of studies, or science in general should unduly influence the decision of when someone should have children.

8

The Time of Your Life

By Paolo Sassone-Corsi, Ph.D.

Paolo Sassone-Corsi, Ph.D., is Donald Bren Professor and director of the Center for Epigenetics and Metabolism at the University of California, Irvine. Before his move to California in 2006, he was director of research at the National Center for Scientific Research (CNRS) in Strasbourg, France. Sassone-Corsi's major interest is concentrated on the mechanisms of signal transduction able to modulate nuclear functions and, in particular gene expression, chromatin remodeling, and epigenetic control. He has received many awards, including the EMBO Gold Medal, the Charles-Leopold Mayer Prize of the Academie des Sciences (France), and the Ipsen Award for Endocrinology (U.S.). He received his Ph.D. in genetics from the University of Naples, Italy.

Editor's Note: The circadian rhythm—the 24-hour cycle of the physiological processes of living beings—is instrumental in determining the sleeping and feeding patterns of all animals, including humans. Clear patterns of brain-wave activity, hormone production, cell regeneration, and other biological activities are linked to this daily cycle. Our author focuses on two relatively new areas of research—circadian genomics and epigenomics—and their potential for advancing medical insight.

EACH MORNING WE WAKE UP from a night of sleep, and each day we eat our regularly timed meals, go through our normal routines, and fall asleep again for another night. This rhythm, so-called circadian—after the Latin words *circa diem* ("about a day")—underlies a wide variety of human physiological functions, including sleep-wake cycles, body temperature, hormone secretion, locomotor activity, and feeding behavior.

A simple look at other organisms reveals that circadian rhythms are remarkably conserved throughout evolution. Whether we consider the cyclic movements of leaves on a plant, the activities of a house cat, or the morning singing of birds, all follow a daily cycle that, being so natural and ancient, generally happens at a subconscious level. Over the past several decades, researchers have described a plethora of cyclic behaviors, metabolic rhythms, and physiological oscillations—all following a circadian pattern. Scientists have observed these behaviors in organisms as different as fungi, insects, unicellular protists, plants, cyanobacteria, vertebrates, and mammals. The rhythms are so wide-ranging that they include both the gravity-driven orientation of the photosynthetic flagellate *Euglena gracilis* and the social behavior of mammals in a group.

Why are circadian rhythms so omnipresent? The answer is straightforward. These biological cycles are based on the most ancient feature of our environment: the astronomical rotation of Earth on its axis, leading to the daylight-darkness cycle—the rhythmic repetition of days and nights.[1,2] This feature has remained immutable over a billion years—although the length of the photoperiod has shortened somewhat over time.[1]

Scientists generally think that living beings have developed by adapting to the daylight-darkness cycle. My personal view is that, in addition to the adaptation process, life has developed *because* of the 24-hour light-dark cycle. Life-forms and their cellular, organismal, and molecular features would have been completely different on a planet with a longer or shorter light-dark cycle. Simple experiments on the small flowering plant *Arabidopsis thaliana* show that its size is reduced by half when subjected to light-dark cycles of 20 hours or 28 hours, corresponding to a planetary rotation that is only one-fifth slower or faster than Earth's.[3]

The role of the circadian clock appears to be so fundamental that, as shown in a number of studies, it has intimate links with the cell cycle.[4] This is nicely illustrated when we consider evolution's role in the process. Indeed, the cell division of a number of unicellular organisms, such as the green alga *Chlamydomonas reinhardtii,* the cyanobacterium *Synechococcus elongates*, and the dinoflagellate *Gonyaulax polyedra*, can be timed by a circadian mechanism. Also, disruption of the clock may have drastic health consequences. In humans, for example, night-shift workers have increased incidence of metabolic disorders.

In the past two decades the knowledge in the field of circadian biology has increased remarkably, such that today it is safe to claim that circadian rhythms represent possibly the ultimate example of systems biology. Some of these fairly recent findings, in my view, have prominently shaped our modern view of the field.

My First Encounter

I attended my very first conference on circadian rhythms more than 20 years ago. I was invited because, while working on the relationship between a messenger important in many biological processes and a gene (cyclic-AMP responsive element modulator, or CREM), my team stumbled on a clever molecular mechanism that allows expression to be cyclic in the pineal gland. Subsequently, we determined that CREM transcriptionally controls the gene encoding the serotonin N-acetyltransferase, an enzyme responsible for the rhythmic synthesis of the hormone melatonin from the pineal gland.

For the most part, the conference was a series of descriptive presentations about measuring circadian oscillations in a wide variety of organisms and physiological settings. Coming from the hard-core field of molecular transcription, I was fascinated by the spectacular variety of biological systems presented and intrigued by the obvious opportunities for mechanistic investigation. Most important, I found (and still find) the self-sustaining nature of circadian rhythmicity thought provoking. The field was on the verge of witnessing a series of conceptual transformations.

What is the evolutionary advantage of circadian clocks? They allow organisms to anticipate daily events (for example, food availability and predator pressure for animals, and sunrise for plants) rather than just reacting to them. Because the measure of time by circadian pacemakers is only approximate, their phase needs to be adjusted daily to stay in synchrony with geophysical time. Self-evident even to nonspecialists, light is the dominant entraining cue for all circadian timekeepers and is consequently considered the most critical *zeitgeber* (German for "time giver") for circadian physiology. In mammals, the anatomical structure that governs circadian rhythms is the suprachiasmatic nucleus (SCN), a small area in the brain consisting of approximately 15,000 neurons localized in the anterior hypothalamus.

For decades scientists have considered this central pacemaker to be the unique circadian clock controlling all daily behavior, metabolism, and physiology.[1,5] SCN neurons are able to self-sustain rhythmicity for weeks even when isolated in a culture dish. Their plasticity is also remarkable: The SCN is reset every day by the light-dark cycle, and thereby undergoes seasonal variations corresponding to the changes in the photoperiod. Thus, SCN neurons are wired to oscillate (to repeatedly move in one direction and back many times), but they receive the light signal through specialized retinal neurons and via the retinal-hypothalamic tract (RHT), thereby insuring their timely adjustment to the changing photoperiod and environment.[5] The SCN's role as the master clock is demonstrated by grafting experiments: Normal SCN grafted into a genetically arrhythmic animal can restore circadian rhythmicity.

Clocks Everywhere!

One discovery that has deeply affected the field of circadian biology during the past 15 years is that oscillators are present in most tissues. The thinking for decades was that the SCN alone directs all circadian body functions, but more recent findings reveal that the liver, spleen, muscle, and other body functions all have their own internal clock. My research team first described this finding in a vertebrate,[6] extending previous observations made in *Drosophila*.[7,8] Soon afterward, this finding was confirmed in mammals.[2,9] In *Drosophila* and zebra fish, light-dark cycles can directly entrain all oscillators, a scenario that is possible only in organisms in which at least some photons reach the internal organs.[6,7]

The evolution of larger and thus opaque organisms necessitated the development of a different, nonphotic (in addition to photic) communication system. In mammals, we see this in the organization of neuroendocrine circuits that convey the timing information from the SCN to the entire organism via direct and indirect signaling pathways. The SCN thereby functions as a master pacemaker, a kind of orchestra director that hierarchically coordinates the subsidiary oscillators located in peripheral tissues.

This notion was further illustrated by an experiment in which fibroblasts (a type of cell that plays a critical role in wound healing) that originated from a mutant mouse (and thereby had a faster clock) took on the rhythm of a host mouse when grafted as a subcutaneous implant.[10] Additional evidence demonstrated the presence of circadian oscillators even in established cell lines: In cultured fibroblasts the endogenous clock system needed a simple serum shock to be resynchronized,[11] while the pacemaker of zebra fish's embryonic cells started ticking upon exposure to a short pulse of light.[2] Together, these findings significantly extended our view of circadian organization at the whole-organism level. They also underscored the fact that circadian-clock functions are not the prerequisite of a relatively small number of SCN neurons, as scientists thought for decades, but instead are common features of most cells.

Yet, more than a decade after these discoveries, some fundamental questions remain unanswered. Specifically, how do SCN neurons communicate and synchronize with the periphery? Are peripheral clocks in different tissues somewhat connected in an SCN-independent manner? Is there any functional feedback from peripheral oscillators back to the SCN?

Solving these points will be highly valuable for biomedical research. As the highly orchestrated network of clocks is based on cascades of signaling pathways, studies by several laboratories focused on understanding how clocks lead to the activation of transcriptional programs that define the unique circadian features of a given tissue. The important surprise came when transcriptional array profiles demonstrated that the clock controls a remarkable fraction of the genome.

Circadian Genomics and Epigenomics

Since the original discovery of the period (*per*) gene in the fly by Ronald Konopka and Seymour Benzer more than 40 years ago, the analysis of clock genes and their relationships and functions has kept an increasing number of researchers busy.[12] At the heart of the molecular network that constitutes the circadian clock are factors involved in turning "on" or "off" transcription organized in feedback loops. This organization ensures cycles of oscillatory gene expression and the control of a remarkable fraction of the genome. Various studies have established that at least 10 percent of all expressed genes in any tissue are under circadian regulation.[13] Additional levels of circadian regulation implicate parallel and intertwined regulatory loops and the control by the cell of clock proteins stability. Moreover, scientists anticipate that tissue-specific transcriptional regulators contribute or intersect with the clock machinery.

The unexpectedly high proportion of circadian transcripts suggests that the clock machinery may direct widespread events of cyclic chromatin remodeling, which is the dynamic modification of chromatin architecture to allow access of condensed genomic DNA. This consequently affects the cycles of transcriptional activation and repression. Remarkably, a recent analysis covering 14 types of mouse tissues identified approximately 10,000 known genes showing circadian oscillations in at least one tissue.

These findings underscore the presence of molecular interplays between the core clockwork, which can be assumed to be common to all tissues, and cell-specific transcriptional systems. Taking into consideration the recent view of the mammalian circadian clock as a transcriptional network, through which the oscillator acquires plasticity and robustness, it is reasonable to speculate that the clock network contributes to physiological responses by intersecting with cell-specific transcriptional pathways.[13]

Considering the thousands of genes regulated in a circadian manner, researchers have questioned how the complex organization of chromatin copes with the task of controlling harmonic oscillations. A number of studies have revealed that several chromatin dynamics contribute to circadian function, rendering specific genomic loci either active (open) or silenced (closed) for transcription.[13] Specifically, we have found that the clock machinery is itself essential for circadian control of chromatin dynamics.

This finding provided a gateway to search for other components of the circadian chromatin complexes.[13] One of these is MLL1, an enzyme implicated in some forms of cancer, that dictates the recruitment to chromatin of the clock machinery thereby targeting circadian genes.[14]

As soon as the first chromatin circadian regulators were identified, the search for the counterbalancing enzymes was open. The discovery that the activity of SIRT1—a longevity-associated enzyme belonging to a family of nicotinamide adenine dinucleotide (NAD+) activated deacetylases—oscillates in a circadian fashion established the first molecular and conceptual link between the circadian clock and metabolism.[15,16] SIRT1 demonstrates an oscillation in activity, impinging back on the circadian clock. The discovery of circadian-directed sirtuin activity spurred hypotheses as to whether metabolites such as NAD+ themselves serve a predominant role in the cellular link between metabolism and the circadian clock.[15,16]

The Metabolic Clock

Intuitively, circadian physiology implies that a considerable fraction of cellular metabolism is cyclic. Also, the analysis of mice mutants for clock proteins has revealed a number of metabolic defects. Indeed, metabolome analyses by mass spectrometry have shown that about 50 percent of all metabolites oscillate in a given tissue. Yet, the question that we have been addressing is as follows: What is the molecular link between clock-driven control and the oscillation of a given metabolite? In this respect, the example of NAD+ is paradigmatic.

Indeed, circadian oscillation of SIRT1 activity suggested that cellular NAD+ levels may oscillate. This is indeed the case, and the way this regulation is achieved is conceptually revealing. The circadian clock controls the expression of the gene encoding nicotinamide phosphoribosyltransferase (NAMPT), a key rate-limiting enzyme in the salvage pathway of NAD+ biosynthesis.[17,18] The clock machinery is recruited to the NAMPT promoter in a time-dependent manner. The oscillatory expression of NAMPT is abolished in mice mutated in clock function, leading to drastically reduced and nonoscillatory levels of NAD+. These results make a compelling case for the existence of an interlocking, classical, transcriptional feedback loop that controls the circadian clock with an enzymatic loop wherein SIRT1 regulates the levels of its own cofactor.[17,18]

More recently, we have questioned whether a nutritional challenge would modify the genomic and metabolomic circadian profile. The nutritional implications of this approach are multiple, especially in a modern society with an endless availability of food. Mice that are fed a high-fat diet (HFD) experience a drastic reprogramming of the circadian clock. Genes that normally would oscillate stop doing so; in addition, many genes whose expression profile is normally noncyclic start to oscillate.[19] The HFD-induced reprogramming pushes the liver to acquire a new circadian homeostasis that implicates genes of the inflammasome and heat-shock response. In this sense, the example of NAMPT and NAD+ is again very revealing. Under HFD, the oscillation of both NAMPT and NAD+ is abolished because the clock machinery cannot be recruited to chromatin. This illustrates that different nutrition strategies directly "talk" to chromatin remodelers, resulting in a reprogramming of genomic functions.[19]

The Next Phase

Mysteries in circadian biology remain. The intrinsic, fundamental role played by the circadian clock in a large array of biological functions illustrates that much more will be unraveled in the upcoming years. Specifically, we predict that the clock will be found to play a key role in the host-pathogen relationship, in the inflammatory response to infection, and in the disturbances caused by tumoral growth. Based on our current knowledge, it will be critical to decipher the role that specific epigenetic regulators have in controlling the circadian epigenome.

Recent findings stress the role of two HDACs of the sirtuin family, SIRT1 and SIRT6, in partitioning the circadian genome in functional subdomains.[20] This partitioning leads to a segregation of cellular metabolism, again underscoring the intimate link with homeostasis.[20] Finally, the role of circadian metabolism in neurons is likely to reveal yet-unexplored regulation pathways that may help us decipher the relationship that the circadian clock has with the sleep-wake cycle.

9

Brain-to-Brain Interfaces

When Reality Meets Science Fiction

By Miguel A. L. Nicolelis, M.D., Ph.D.

Miguel Nicolelis, M.D., Ph.D., is the Duke School of Medicine Professor of Neuroscience; professor of neurobiology, biomedical engineering, and psychology and neuroscience; and founder of the Duke University Center for Neuroengineering. He is founder and scientific director of the Edmond and Lily Safra International Institute of Neuroscience of Natal. Nicolelis is also founder of the Walk Again Project, an international consortium of scientists and engineers dedicated to the development of an exoskeleton device to assist severely paralyzed patients in regaining full-body mobility. Nicolelis, who proposed and demonstrated that animals and human subjects can utilize their electrical brain activity to directly control neuroprosthetic devices via brain-machine interfaces, is a member of the French and Brazilian Academies of Science, holds three U.S. patents, and has been published in *Nature, Science,* and *Scientific American.*

Editor's Note: Every memory that we have, act that we perform, and feeling that we experience creates brainstorms—interactions of millions of cells that produce electrical signals. Neuroscientists are now able to record those signals, extract the kind of motor commands that the brain is about to produce, and communicate the commands to machines that can understand them and facilitate movement in the human body. Research in this area has the potential to help paraplegics and others suffering from spinal-cord injuries to control machines with their thoughts and to bolster their ability to get around.

WHEN ASTRONAUT FRANK POOLE, deputy commander of the USS *Discovery*, miraculously wakes after his 1,000-year slumber in deep space, the first thing he notices is that the medical personnel who kindly tend to him rarely speak to one another. In this scene of Arthur C. Clarke's science-fiction novel *3001: The Final Odyssey,* Poole's attendants utter words—more like whispers—only when they have to communicate with their stunned, severely obsolete patient. Their inaudible speech may be absolutely natural for these men and women of the 31st century. However, for someone whose last enduring memories include an upsetting (and audible) argument with a temperamental supercomputer named HAL while orbiting Jupiter in 2001, this form of communication certainly feels spooky.

In time, with a mix of trepidation and awe, Poole learns that by 3001, thanks to the invention of the so-called Braincap, virtually every Earth inhabitant had acquired the ability to communicate directly with computers—and even with other humans—simply by thinking! Mankind had mastered an amazing range of additional tricks, including uploading gargantuan amounts of information and knowledge directly into brain circuits or, conversely, downloading an entire life's history into some sort of perpetual storage medium.

When *3001: The Final Odyssey* was first published in 1997, Clarke surprised many of his readers by dedicating a significant chunk of his plot to Braincaps and their impact on daily human behavior a thousand years from now. Judging by some of the reviews that followed the book's publication, many readers and critics thought that Clarke had gone too far, even considering his impressive record as a futurist.

Yet Clarke's vindication came much sooner than his critics would have expected. After all, just a couple of years after the publication of his book, American and European laboratories started reporting on pioneering experiments that employed real-time links connecting living brain tissue with artificial devices. Such brain-machine interfaces (BMIs), as this new paradigm was named, allowed either animals or severely disabled patients to use the brain's electrical activity to control the movements of artificial devices in order to execute simple tasks.

Early Advances

For instance, in 1999, John Chapin's laboratory at Hahnemann University in Philadelphia and my own laboratory at Duke University collaborated in the first experimental demonstrations of a brain-machine interface in animals. In these experiments, rats learned to use the combined electrical activity of a handful of cortical neurons to move a robotic arm in order to obtain a water reward. Around the same time, Niels Birbaumer at the University of Tübingen in Germany reported how completely paralyzed patients learned to use brain-derived signals (recorded though a classic method known as electroencephalography, or EEG) to write messages on a computer screen. Even in its initial version, this brain-computer interface was the only way for these locked-in patients to communicate with the external world. It was an early indication of BMIs' significant potential as new rehabilitation tools.

In subsequent years, further animal experiments with BMIs indicated that monkeys could learn to employ the combined electrical activity of hundreds of their cortical neurons to move multiple degree-of-freedom robotic arms, entire humanoid robots, and even avatar limbs and bodies without the need for any overt movement of their own bodies. Soon, initial clinical studies also reported that patients could rely on BMIs to control the movements of computer cursors and robotic arms.

As the BMI field rose to the forefront of modern neuroscience, the possibility of establishing a bidirectional dialogue between brains and artificial devices was also realized. In 2011, through a technique called cortical electrical microstimulation, my laboratory was able to deliver simple "tactile messages" directly into the brains of monkeys. Every time one of our monkeys used its brain activity to move a virtual hand to scan the surface of a virtual sphere, a simple electrical wave, proportional to the virtual texture of the touched object, was immediately delivered to the animal's primary somatosensory cortex, an area known to be fundamental for the definition of one's tactile perceptions. After a few weeks of training, by taking advantage of this direct and continuous inflow of tactile information into their brains,

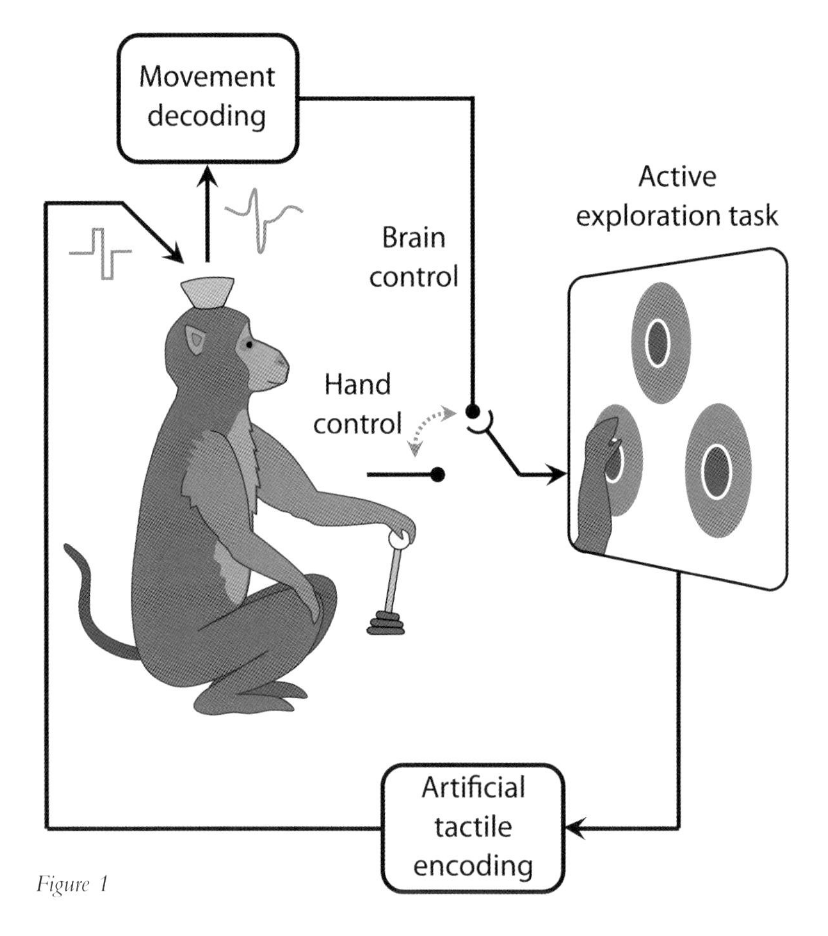

Figure 1

a pair of monkeys became capable of discriminating the fine texture of the virtual objects by using their virtual hands, as if they were using their own biological fingertips. We called this new paradigm a brain-machine-brain interface (BMBI, Figure 1).

A Major Breakthrough

In 2009, as a direct result of this auspicious first decade of BMI research, the Duke University Center for Neuroengineering and the Edmond and Lily Safra International Institute of Neuroscience of Natal (ELS-IINN, in Brazil) jointly created a nonprofit research consortium called the Walk Again Project. By the end of 2012, the Walk Again Project received a grant from the Brazilian government to assemble a large international research team of roboticists, neuroscientists, engineers, and computer scientists. This international team joined with a Brazilian multidisciplinary rehabilitation team, composed of physicians, psychologists, and physical therapists, to take on a very ambitious project: designing and implementing the first bipedal robotic exoskeleton whose movements could be controlled directly by human-brain activity. The central goal of the first phase of project was to allow paraplegic patients suffering from severe spinal- cord lesions to use their EEG activity to control the exoskeleton's leg movements (Figure 2) and, in so doing, regain lower-limb mobility. In addition to restoring basic locomotion behaviors, the exoskeleton would be the first in its class to provide continuous sensory feedback to the user in the form of artificial tactile and proprioceptive signals.In December 2013, a group of eight

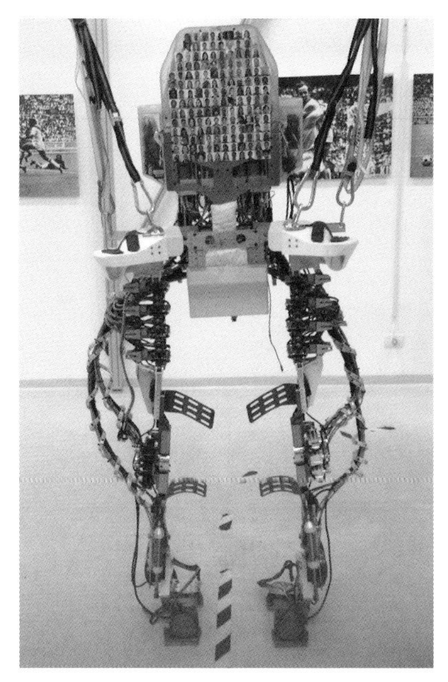

Figure 2

patients suffering from complete and incomplete spinal- cord lesions started the training process required for achieving proficiency in controlling a brain-controlled robotic exoskeleton. Four months later, all eight were capable of commanding the exoskeleton with their brain activity alone, and all had regained the sensation of walking in a laboratory setting. The feeling of walking again was even more realistic in these patients because of the addition of two innovative technologies in the design of our exoskeleton. The first was a new type of artificial tactile sensor known as artificial skin, developed by Gordon Cheng at the Technical University Munich. These sensors were distributed across key locations of the exo's legs and feet to detect the device's movements and contact with the ground. The second was an ingenious haptic display, created by Hannes Bleuler's laboratory at the École Polytechnique Fédérale de Lausanne (EPFL) in Switzerland, that allowed the tactile feedback signals generated by the arrays of the artificial sensors to be delivered to the skin of a patient's forearms. For the haptic display to work properly, patients had to wear a special shirt containing a linear array of small vibromechanical elements in the distal half of each sleeve while walking with the brain-controlled exoskeleton.

The World Cup Demonstration

To celebrate a major first step toward the development of a new generation of neuroprosthetic devices, one of our patients, Juliano Pinto, who is paralyzed from the mid-chest down, was invited to help our team demonstrate our exoskeleton's enormous potential before the opening match of the 2014 FIFA World Cup in Brazil on June 12. For the first time in history, a human subject showed that a brain-controlled exoskeleton could be used to initiate the kicking of a soccer ball. The demonstration was witnessed by 70,000 fans at the Itaquerão stadium and an estimated 1 billion people watching on TV. Seconds after executing this historic kick, Juliano reported to us that he clearly felt his leg moving in the air during the moment at which the exo's foot made contact with the surface of the ball. According to Juliano's perception, it was his own body, not the exo, that executed the kick. This was a stunning development.

The effort and complexity required to pull off that World Cup demonstration well exemplifies the current state of the art of BMIs. Over the last 15 years, since our initial study with rats launched the field, progress has been steady. And although the case has been made that BMIs offer concrete hope for the future development of a variety of new neurorehabilitation tools, we are still a few years away from being able to produce neuroprosthetic devices that patients can routinely use outside well-controlled laboratory conditions. Certainly, at this point, we are very far from the Braincaps imagined by science-fiction writers like Clarke. Indeed, we may never get there at all.

Enterprising Explorations

Despite the uncertainty, recent experiments combining BMIs with cortical electrical microstimulation effectively open the doors for more daring adventures. Indeed, I could almost bet that Clarke himself would have enjoyed the opportunity to be present when Miguel Pais-Vieira, a Portuguese postdoctoral fellow in my lab, demonstrated the operation of the first brain-to-brain interface designed to link two animals' brains directly (Figure 3). First proposed in my 2011 book *Beyond Boundaries: The New Neuroscience*

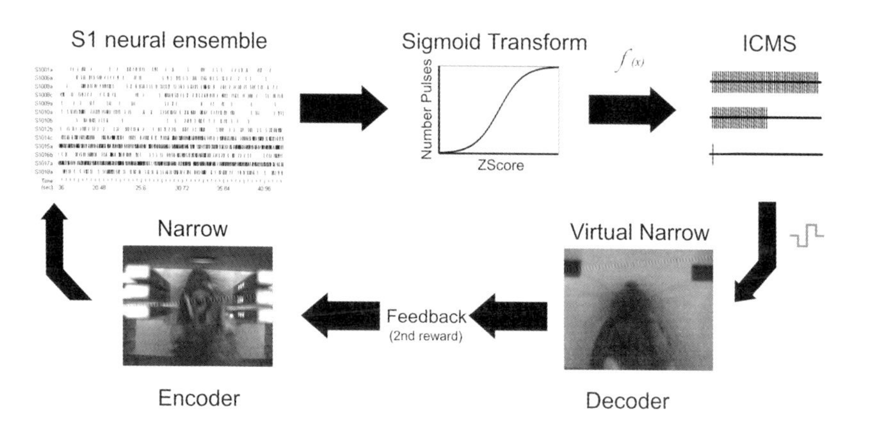

Figure 3

of Connecting Brains with Machines—and How It Will Change Our Lives, our brain-to-brain interface (BTBI) paradigm, reported in 2013, allowed a pair of rats to transmit and receive rudimentary mental sensorimotor messages.

In one of the published experiments, the first rat of the pair, known as the encoder, was trained to use its facial whiskers to determine the diameter of a computer-controlled aperture placed inside a behavioral box. From trial to trial, the aperture could assume two distinct diameters, classified as narrow (X mm) or wide (Y mm). The encoder's job was to use its facial whiskers to correctly judge the aperture's diameter and then indicate its value by placing its snout in one of two holes located in a nearby chamber. If the encoder nose-poked in the hole corresponding to the correct aperture diameter, it received a water reward.

As the encoder used its facial hair to evaluate the opening's diameter, electrical activity recorded from neurons located in its somatosensory cortex was combined and transmitted, via cortical electrical microstimulation, to the brain of a second rat, the decoder, located in a different behavioral box. The decoder had no access to an aperture, so its facial whiskers were useless in solving the task and getting water. Yet to receive such a reward, the decoder also had to indicate, by nose poking, the diameter of the aperture touched by the encoder. To do that, the decoder had to rely solely on the simple neural message being transmitted to its brain by electrical microstimulation.

After a bit of training, decoder rats became capable of using our brain-to-brain interface to successfully perform this task way above chance level. This indicated that the brain of a decoder rat could make sense of the messages broadcasted by its associated encoder rat. Interestingly, since the encoder received an extra reward allotment every time a decoder was able to correctly indicate the aperture's diameter, encoder rats adapted their behavior and cortical activity to make it easier for their counterpart to complete the task, particularly after the latter committed a series of trial errors.

That further suggested to us that these rat dyads had established a new form of communication, despite the fact that neither animal was aware of its counterpart's existence. As an extra proof of the effectiveness of this BTBI, we repeated these tactile-discrimination experiments by using an encoder rat placed in a laboratory in the ELS-IINN, in Natal, Brazil, while the decoder rat performed its trick in my lab at Duke University, in the U.S. Despite the distance and the use of an average Internet connection, the brain-to-brain interface worked as well as it did when the two animals were in the same laboratory.

In a final test of our BTBI, decoder rats used neuronal signals provided by the motor cortex of encoder rats to choose which of two levers to press, without ever seeing the visual cues that instructed the encoders to make the same decision in the first place. In other words, the brain-to-brain connection between the encoder and decoder rats allowed the latter to correctly make a motor decision based on visual cues experienced only by its encoder partner.

During the past year, two other laboratories published studies involving brain-to-brain architectures. Moreover, a press release from a group at the University of Washington indicated that the group had established a functional link between human subjects' brains by combining two noninvasive techniques: EEG to record brain activity in the first subject (encoder) and transcranial magnetic stimulation (TMS) to deliver an EEG-triggered signal to the second subject's (decoder's) brain. Since the group has not yet published a full scientific report, it is difficult to evaluate what was really achieved. If anything, the limited description (and video clip) provided in the press release did not fully support the claim that a true functional communication between human brains occurred. This is because, essentially, the encoder's EEG activity was simply used to trigger a magnetic stimulus in the decoder's motor cortex.

As expected, every time this magnetic stimulation was delivered, the decoder subject produced an involuntary body movement. Yet the decoder was unable to actually participate in the decision to create the movement. As such, I do not see how two brains shared a true message in this paradigm. On the other hand, the potential methodology for doing so was unveiled, and that was certainly enough to cause a significant media and public response.

A New Type of Computational Architecture?

In my lab at Duke, we continue to experiment with animal BTBIs. We are currently investigating what kinds of social behaviors and global patterns of neuronal activity emerge when groups of animal brains are allowed to collaborate directly, through the employment of different types of brain-to-brain interfaces. I like to refer to these systems as Brainets. So far we have tested Brainets formed by either four rats or three monkeys. The central task of each of these animal Brainets is to optimize the combination of neuronal activity, sampled from multiple brains simultaneously, into a supranervous system that is responsible for attaining a common behavioral goal, such as identifying a complex tactile pattern or moving an elaborate virtual limb. The results of these studies, which are currently under review for publication, mostly focus on how BTBIs can enhance social interactions between animals and whether Brainets could operate as a new type of computational architecture, like some sort of non-Turing biological computer. In addition, these experimental paradigms allow one to study whether, in a still-distant future, artificial interfaces like these may be used to functionally reconnect brain areas where communication may have been disrupted by brain damage, such as that produced by strokes or other neurological disorders.

Right now this latter proposition may sound farfetched. However, seeking such a path has become a hallmark of our laboratory during the past decade. During this period, we have successfully translated similarly abstract basic-science ideas into potential new therapies for untreatable epilepsy, Parkinson's disease, and disabling paralysis. All of these therapies are currently undergoing clinical testing worldwide.

As exciting as these animal research projects are, none of them come close to competing with the fictional wonders of Clarke's Braincaps. But that may not be so bad after all. For starters, nobody would ever consider it ethically or medically acceptable to implant nanotubes or other types of electrodes in healthy human subjects for the purpose of testing a BTBI, as suggested by Clarke. But even if, years or decades from now, better noninvasive technology enables us to record large-scale brain activity in real- time, at the millisecond scale, and then another efficient, noninvasive method might be used to deliver brain-derived messages to another human brain, it is highly unlikely that such a BTBI would lead to the emergence of a fluid and efficient form of human communication, as long as we rely on digital computers to mediate this task.

Nor do I believe that there will be a day in which Braincap-like technologies will allow us to upload vast and complex information packets—like a new language or a large amount of scientific knowledge, as Clarke describes in his book—into our brains, or to download all our memories or personal experiences into some sort of digital storage media. Apart from tasks such as motor control for which BMIs can become very useful, mimicking higher-order brain functions, such as knowledge acquisition, memory storage, performance of cognitive tasks, and even consciousness, may be beyond the reach of binary logic, the basis from which all digital computers operate, no matter how simple or elaborate. An interesting corollary of this view is that we need not worry about the forecast that, in the near future, a "really smart" digital computer/machine will supplant human nature or intelligence. In all likelihood, this day will never come because, in a more-than-convenient arrangement, our most intimate neural riddles seem to have been properly copyright-protected by the very evolutionary history that generated our brains, as well as the very complex emergent properties that make it tick. As such, neither evolution nor neurobiological complexity can be effectively simulated by digital computers and their limited logic.

In the end, this may not be so bad. Like Commander Poole, as much as I would love to take advantage of a brand-new Braincap—minus the nanotubes—to learn a few new intellectual tricks in a hurry, from the perspective of someone living in the early 21st century (not the 31st), it is very difficult to imagine that any of us at this juncture in our history would, in good faith, feel comfortable in surrendering our final frontier of individual privacy, knowing that there is a chance, no matter how insignificant it may be, that an uninvited snoop may, nevertheless, want to take a peek.

10

With a Little Help from Our Friends

How the Brain Processes Empathy

By Peggy Mason, Ph.D.

Peggy Mason, Ph.D., is a professor of neurobiology at the University of Chicago and the author of *Medical Neurobiology* (Oxford University Press, 2011). Dr. Mason offers an open online course, "Understanding the Brain: The Neurobiology of Everyday Life," through Coursera (https://www.coursera.org/course/neurobio). She also maintains a blog at http://thebrainissocool.com/. For more than 20 years, Dr. Mason's research was focused on the cellular mechanisms of pain modulation. In the last several years, she has turned her energies to the biology of empathy and prosocial behavior. Originally from the Washington, D.C. area, Dr. Mason received her bachelor of arts degree in biology in 1983 and her Ph.D. in neuroscience in 1987, both from Harvard University. After postdoctoral work at the University of California-San Francisco, she joined the faculty at the University of Chicago in 1992. A lively discussion of her empathic helping work can be found at reddit.com/r/science/comments/23o5w4/science_ama_series_hi_im_peggy_mason_i_study.

Editor's Note: Why are certain individuals born with a brain that is wired to help others? What daily habits or life experiences reinforce compassion but also selfishness, narcissism, and psychopathy? Social neuroscience models have assumed that people simply rely on their own emotions as a reference for empathy, but recent studies suggest neurobiological underpinnings for how the brain processes empathy. A better understanding of these processes, says the author, could lead to more social cohesion and less antisocial harm in society.

"No man is an island entire of itself." — JOHN DONNE

JOHN DONNE'S SENTIMENT THAT THE SELF extends beyond our individual boundaries applies to virtually every animal. Mammals in particular depend on others for growth, development, effective foraging, safety, and ultimately survival. The young are born helpless and depend on their mother for nutrition, immune defense, thermoregulation, and protection from predation. If a mother (or father in some birds and a few mammals) fails to recognize that her young are cold, hungry, or exposed to danger, it's likely that the offspring will die. A mother's ability to react appropriately to her newborn's needs is the difference between the newborn living and carrying on its parents' genes or dying and reaching an evolutionary dead end.

Young mammals depend on their mothers for far more than nuts-and-bolts survival. American psychologist Harry Harlow dramatically illustrated this point in the late 1950s through an experiment that involved separating monkeys from their mothers and offering two mother substitutes, one made of cloth and the other of wire.[1] Even when the wire mother was the baby's only source of milk, the young spent 10 times as much time with the cloth (nonnutritive) mother than with the wire (nutritive) mother. The experiment revealed that mammalian offspring crave a mother's touch, which is better approximated by cloth than wire, and demonstrated that a primate mother provides her newborns with far more than nutrition. Indeed, mammals raised with adequate food, warmth, and protection but with diminished social contact became fearful, anxious adults with impaired social and parenting skills.[2,3]

Whereas an empathic understanding by the mammalian mother of her young's condition is critical for survival and, in turn, for evolutionary success, the utility of a social bond between individuals extends into adulthood and encompasses more than just the mother-to-offspring relationship. Darwin intuited that an individual should "extend his social instincts and sympathies to all the members of the same nation, though personally unknown to him."[4] If we take "nation" to mean tribe or herd or social group, Darwin is clearly stating that social cohesion, borne of mutually directed feelings, facilitates survival of the group and its constituent individuals. In essence, sociality benefits individual adult survival by providing protection and by increasing opportunities to feed, mate, and successfully raise offspring to reproductive age.

Membership in a social group brings benefits that scale with the ability of the group to work together. Thus social cohesion, more than simple sociality, most powerfully promotes survival. William James considered that "a man's Self is the sum total of all that he CAN call his, not only his body and his psychic powers, but his ... friends."[5] James's idea of an extended self depends on individuals reacting to the fortunes of others as they would if the same fortune or misfortune befell oneself. To the extent that our mood soars at a friend's triumph as it does upon our own triumph or plummets in reaction to harm befalling a friend, that friend is part of a Jamesian extended self. Social cohesion increases as more group members consider more other group members as part of their extended selves.

To Act, or Not to Act

At the core of social cohesion among mammals is the communication of affective or emotional states between individuals. When individuals respond to others' emotional states as if they were their own, the result is a bond, thereby building social cohesion within the larger group. The communication of affect or emotion between individuals is empathy. Defined in this way, empathy is an umbrella term that includes a large range of interactions in which an emotional or affective response is elicited by the emotional or affective state of another individual. Moreover, according to this definition, empathy is neutral in that responding to another's affective state, mood, or emotion does not constrain the actions taken, if any, as a result. We may hope that an individual reacts with helping behavior to a member of his or her own species in distress—the social instincts and sympathies Darwin suggested. Yet, inaction and even targeted cruelty aimed at exacerbating a victim's distress are also possible reactions.

The perception-behavior link, an automatic function that links our behavior to the behavior of another, is critical to affective communication between two individuals. Many of us are familiar with the phenomenon of adopting the physical stance of a person or people with whom we are talking; soon after one member of a group crosses his or her arms, another person does the same.

Simply viewing another individual's actions increases the probability that the viewer will perform the same actions—even if the individuals are strangers.[6] Similarly, people in conversation with each other modify their fundamental speech frequencies to more closely match each other.[7] These social adjustments make the actions of two interacting people more similar to each other and serve as an affiliative signal, or a kind of social glue. Passing a person who cheerfully smiles at us makes us more likely to smile. We don't reason through this process; it just happens.

Actions are not only the readout of affect. They also influence affect—the interaction between emotions and outward expressions is two-way.[8] In other words, just as our emotions lead to actions, our motor actions are "re-experienced" as affect. Affect and emotions are expressed through voluntary muscles responsible for posture, facial expression, breathing, and gaze, as well as autonomic processes such as a rise in heart rate or perspiring, blanching, and blushing. Facial expressions' influence upon emotional experience is particularly strong in humans. People report emotions commensurate with artificially arranged facial expressions.[9] Feeling happy makes us smile and smiling can make us feel happy, or at least happier. When you're feeling good and laughing with friends, just try to feel angry or sad. As long as you keep your face in a smile or laugh, feeling an incongruent emotion is nearly impossible. Deriving emotion from action, often termed embodied emotion, is the essence of the Stanislavski system of method acting in which the affects that emerge from movements provide the emotive force of a performance.

The links between perception and action and between action and affect set up a cascade whereby one person's perception by another's actions ultimately results in the first person feeling the second one's mood. This cascade results in matching affects.[10] The affect the viewer experiences is vicarious in nature, "caught" from the other individual. The process by which one individual catches the affect or emotion of another is called "emotional contagion," and it is a fundamental building block of more complex forms of empathy.

Aid and Abetment

Emotional contagion is required but not sufficient to elicit empathically motivated helping. In humans, personal distress must be suppressed in order to move from emotional contagion to helping, to choose action over immobility and panic. High levels of personal distress are detrimental to helping.[11] Suppressing personal distress allows someone to focus on the other over the self and leads to empathic concern, an other-oriented emotional response elicited by and congruent with the welfare of an individual in distress. The response's congruence with the welfare of the other precludes antisocial actions so that the action taken by someone feeling empathic concern is always prosocial in nature.

By helping a distressed individual, a helper resolves not only the distressed individual's predicament but also his or her own uncomfortable affective state, providing an internal reward.[12] Thus, helping dissipates the distress of both the helper and the beneficiary. That the helper benefits does not diminish the prosocial action or its effect. The empathy-helping connection is so effective precisely because "empathy gives individuals an emotional stake in the welfare of others."

What Nonhumans Reveal

Empathy is an internal experience. The feeling of empathy may drive behavior such as a facial expression. In humans, empathy may even drive speech. Nonhuman animals, however, do not have the control over, and variety of, facial expressions that humans possess. Therefore, probing a nonhuman animal for the internal feeling of empathy has proved challenging.

Researchers have taken two basic approaches. One has been to test for emotional contagion. A typical experiment using rodents tests the influence of one rodent's expression of either fear or pain on another rodent's behavior. The second approach asks whether, given the opportunity, animals will engage in prosocial behavior, such as sharing food (in primates) or working for the relief of another from foot shock or confinement (in rodents). Mounting evidence suggests that emotional contagion and prosocial behavior are present in nonhuman primates and rodents and likely are widespread among mammals.

Preverbal humans and nonhuman primates show prosocial behavior. For example, a useful paradigm to test for helping behavior is to place an object out of the reach of an experimenter but within reach of the test subject. The subject watches the experimenter try to reach for the object unsuccessfully. The question is whether the subject will hand the object to the experimenter even though the subject gains no reward by doing so. In one study using this paradigm, chimps handed the object to the experimenter in about 40 percent of the trials; human infants helped in about 60 percent of trials.[13] The proportions of children (60 percent) and chimps (50 percent) that helped at least once in 10 trials were similar. The subsets of children and chimps that had helped were then tested on the same task but with physical obstacles placed between them and the object, adding to the cost of helping. Just over half of the subjects of both species helped even when helping required significant effort.

Since emotional contagion is automatic, from perception to action to embodied emotional cascade, it is not surprising that chimps and other primates also appear to experience empathy.[12] Moreover, because the pathways involved in linking perception to action to embodied emotional cascade are shared across mammals and the perception-action model for empathy does not depend on conscious deliberation or higher cognition, there is no reason to expect that empathy and prosocial behavior are exclusive to primates.[12,14] Indeed, emotional contagion has been documented in a number of mammalian species, including rodents.[15] A mouse that views another mouse experiencing foot shock, for example, shows fear by freezing in place.[16] In another example, pairs of mice that receive a noxious stimulus show more pain behavior than a single mouse that receives the same noxious stimulus; this finding has been interpreted as emotional contagion of pain.[17]

Overcome by Caring

Several years ago, we designed a behavioral helping test for rats.[18] In this test, one rat is restrained in a plastic tube located in the center of an arena while a second rat is free to roam in the arena. The restrainer door can be opened only from the outside—only by the free rat. Within a few sessions, most rats begin to open the door consistently, releasing the other rat. By the final session, most rats open the restrainer door within just minutes. The fact that rats opened the restrainer door repeatedly and consistently is remarkable in light of rats' strong preference to remain close to walls and avoid open areas. The motivation to approach the trapped rat in the arena center evidently is sufficient to overcome rats' natural avoidance of open space.

The rats' helping actions are remarkable for another reason. We would expect that emotional contagion would lead the free rat to experience at least some of the distress felt by the trapped rat. The most common reaction of a rat to personal distress is freezing or immobility such as that which occurs in response to foot shock. Yet rats in the helping test do not freeze. Instead they act intentionally to open the restrainer door.[18] This behavior suggests that the helper rat recognizes that its distress is vicarious in origin. In other words, the rat is able to attribute personally felt distress to the trapped rat's condition and distinguish that from its own condition. Such recognition of the distinction between self and other is unexpected in a rat.

The helping behavior test is not the first scientific demonstration of helping, but it is the first tractable paradigm for studying prosocial behavior in a mammal. Already, the test has been used to demonstrate that helping is socially selective.[19] Rats help a stranger rat but only if that rat is from a familiar strain. In other words, an albino rat that has never before seen a black-hooded rat will not help it. However, an albino rat that has lived with a black-hooded rat will open the restrainer door for black-hooded strangers. Remarkably, albino rats raised since birth with black-hooded rats do not help other albino rats, although they do help black-hooded strangers. This suggests that rats do not inherit genetic instructions to help others of their own kind. Instead, they learn which individuals to help from their social environment. The test result tells us that environmental experiences trump genetics when it comes to targeting helping, resolving one piece of the nature-nurture debate. Moreover, because the fostering-from-birth experiment is easy to perform in rodents but would be impossible in most other mammals, this result shows the power of an experimental model for prosocial behavior in rodents.

The finding that rats help strangers of a familiar type but not strangers of an unfamiliar kind may appear, at first glance, to suggest a biological basis for a social bias, a kind of "strainism." However, the results are more consistent with a biological basis for "groupism" through social experience. Humans readily form strong affiliations to groups that are based on "minimal-group" criteria such as an arbitrary assignment to one of two meaningless markers (e.g., red or blue wrist bands).[20] The finding that rats raised without experience with their own strain do not help strangers of their own strain demonstrates that group affiliation, with respect to helping, is fluid, based on experience, and not genetically determined.

Motivation to Help

We still need to understand more about the rat's motivation to help another in distress and to discover the underlying brain mechanisms that support helping behavior. While the motivation for prosocial behavior looks like empathic concern, a rat may open a restrainer door for other reasons. One commonly raised possibility is that the rat finds some part of the trapped rat's behavior so aversive that it opens the door to terminate this aversive experience.

Since rats do not open restrainer doors for rats from unfamiliar strains, including strangers from their own strain if they were fostered with a different one, escaping aversion is a possible but unlikely motivation for door-opening in the helping behavior test.[19] Nonetheless, empathic concern must start with an individual showing distress. Because an individual's demonstration of distress is as critical to empathic concern as another's noticing and responding to that distress, biology has left little to chance. Crying works—babies and others get attention when they need help. Facial expressions and posture also work because they are universal, with commonalities across populations and across species. Conversely, in the absence of an individual displaying distress, nonhuman animals and young humans are never moved to "help." Some degree of attention-getting distress is necessary to elicit empathic concern.

A second common theory is that rats may be motivated by a desire to interact socially with the trapped rat. Social reward is a fundamental underpinning of social behavior, and all rodent behavior involving more than one individual is, at the very least, influenced by social reward. Rats will opt to be together when given the opportunity. Using a modification of the helping behavior test in which rats were repeatedly retrapped, a free rat that could not release the trapped rat opted for physical proximity.[21] In contrast, when given the chance, rats continue to open the door for a trapped cagemate even when subsequent interactions are prevented.[18] This finding suggests that the opportunity to play or interact with the trapped rat is not a requirement for prosocial behavior.

More Than a Feeling

Not all humans show empathy or express helping behavior. We aren't alone: About 25 percent of the rats that we have tested in standard conditions do not exhibit helping behavior. The predominant reason for not helping appears to be an excessive amount of personal anxiety. Similarly, bonobo apes who show more anxiety (measured by how much they scratch themselves) and take longer to recover from a stressful event show less consolation behavior toward other bonobos in distress.[22] This finding dovetails beautifully with research in humans suggesting that in order to use empathy for helping or caring, an individual must overcome personal distress, a process typically termed self- or down-regulation. People with a specific genetic variation who show greater social anxiety also demonstrate less helping behavior.[23] This finding suggests that rather than lacking empathy, many individuals who do not help may be unable to suppress the anxiety associated with catching another's feeling of distress.

In professions that involve repeated exposure to human suffering, such as medicine, strong down-regulation is highly adaptive in counteracting the development of burnout. Physicians have a down-regulated response to noxious events that are common in medical practice. For example, an image of a needle stick evokes a lower assessment of pain intensity and unpleasantness by physicians than controls.[24] Finally, human psychopaths appear to lack empathy and exhibit a callous disregard for others' suffering. Whether psychopathic individuals exist in other mammalian species is an unanswered question.

Researchers are beginning to elucidate the brain circuits that support empathy and empathic concern. A particularly instructive approach has been to compare brain activation in humans, using functional magnetic resonance imaging (fMRI), to compare when an emotion is experienced by the self versus when it is experienced by another. When one person views another in pain, the activated brain areas are similar and overlapping, but not identical, to those activated by a personal experience of pain.[25] The overlapping regions of activation evoked by self- and other-pain can breed confusion so that an individual experiences another's distress as their own. Conflation of the distress originating with the self and the other may explain why vicarious distress can be as immobilizing as personal distress. It appears that the prefrontal cortex allows us to make the distinction between ourselves and others by promoting down-regulation.[25] For example, the medial and dorsolateral prefrontal cortex was activated as physician acupuncturists viewed images of needle insertions, an activation that was not observed in control subjects.[26] Moreover, the degree of activation in the prefrontal cortex was inversely correlated to the ratings of pain intensity made by the subjects so that those with the greatest prefrontal activation judged the needle insertions with the lowest pain ratings.

These studies show that viewing others' pain engages ascending affective pathways, while top-down regulation arising from the prefrontal cortex is critical to stemming personal distress so that empathy can serve as a call for action. Nonhuman animals are likely to employ similar brain circuits. Indeed, fear contagion in mice appears to require the anterior cingulate cortex as well.[16,25] A full elucidation of the similarities and differences in brain circuits involved in empathy and down-regulation between humans and other mammals is an exciting challenge for the future.

11

The Brain-Games Conundrum

Does Cognitive Training Really Sharpen the Mind?

By Walter R. Boot, Ph.D., and Arthur F. Kramer, Ph.D.

Walter R. Boot, Ph.D., is an associate professor of psychology at Florida State University. Boot has published extensively on the topic of technology-based interventions involving digital games, and is one of six principal investigators of the Center for Research and Education on Aging and Technology Enhancement. He conducts studies of aging road users; specifically examining countermeasures to protect older adults as they navigate roadways as drivers, cyclists, and pedestrians. Boot received his Ph.D. in psychology from the University of Illinois at Urbana-Champaign in 2007.

Arthur F. Kramer, Ph.D., is the director of the Beckman Institute for Advanced Science & Technology and the Swanlund Chair and professor of psychology and neuroscience at the University of Illinois. A major focus of his lab's recent research is the understanding and enhancement of cognitive and neural plasticity across the lifespan. He is a former associate editor of *Perception and Psychophysics* and is currently a member of six editorial boards. Kramer, who received his Ph.D. in cognitive/experimental psychology from the University of Illinois in 1984, is also a fellow of the American Psychological Association, American Psychological Society, a former member of the executive committee of the International Society of Attention and Performance, and a recipient of an NIH Ten Year MERIT Award. His research has been featured in the *New York Times, Wall Street Journal, Washington Post,* and *Chicago Tribune,* and on *CBS Evening News, Today Show, NPR* and *Saturday Night Live.*

Editor's Note: Few topics in the world of neuroscience evoke as much debate as the effectiveness of cognitive training. Do you misplace your keys regularly? Forget appointments? Have trouble remembering names? No worries. A host of companies promise to "train" your brain with games designed to stave off mental decline. Regardless of their effectiveness, their advertising has convinced tens of thousands of people to open their wallets. As our authors review the research on cognitive-training products, they expose the science surrounding the benefits of brain games as sketchy at best.

THE NUMBERS ARE STAGGERING. Brain-training products are a billion-dollar industry whose revenues are predicted to surpass $6 billion by 2020.[1] One of the more popular brain-training programs, Lumosity, recently reached the milestone of 50 million members,[2] likely in part due to an advertising campaign that spanned radio, television, and the Internet. Nintendo's *Brain Age* has sold millions of copies and is among the best-selling Nintendo DS games of all time.[3] These statistics suggest a belief that brain training produces meaningful benefits, and this belief does not appear to be restricted to individual consumers. The fact that health-insurance companies have begun making brain-training products available to their clients suggests a perception in the health-care industry that the products work.

But the issue of what does and doesn't work is complex. The basic assumption behind almost all commercial brain-training programs is that practicing one or more tasks leads to improved performance of other, untrained tasks. The programs often present individuals with a series of simple games that might require the player to remember the properties of briefly presented pictures, to keep track of multiple moving objects, to recognize complex patterns, or to rapidly detect the presence of target objects in the visual periphery. With practice, players become faster and more accurate at performing these tasks. These products would be of little value if players improved *only* on the trained games, however. The critical question is whether *transfer of training* occurs. Does extended practice of the trained games result in general perceptual and cognitive improvements that boost performance of meaningful, real-life tasks such as driving, remembering names and faces, and keeping track of finances?

Where It Began

Psychologists have systematically studied the issue of transfer of training for over a century. The early work of American psychologist Edward Thorndike (1874–1949) set the stage for much of the research that followed. Thorndike, a professor at Columbia University's Teachers College, conducted a series of influential studies in which participants practiced one task (for example, estimating the area of a rectangle) repetitively for an extended period of time. After participants demonstrated improvement, he would ask them to perform a different, so-called transfer task (such as estimating the area of a triangle). Thorndike consistently observed large gains on practiced tasks, but these gains were weakly (if at all) associated with improved performance on transfer tasks. Thorndike also tested the idea that training in Latin would result in a more disciplined mind, which would improve performance in a variety of other subjects. He observed no such advantage.

Thorndike's research led him to conclude that transfer of training occurred only if the practiced task and the transfer tasks shared "identical elements,"[4] and that the elements of most tasks are different enough from those of other tasks that transfer of training was rare. He believed that "the mind is so specialized into a multitude of independent capacities that we alter human nature only in small spots, and any special school training has a much narrower influence upon the mind as a whole than has commonly been supposed."[5]

Thorndike's conclusions are consistent with modern theories proposing that practice results in cognitive adaptations that develop over time and are specific to the practiced task,[6] as well as with theories of learning that link improved task performance to the retrieval from memory of specific instances of the same task encountered previously.[7] The implication of Thorndike's empirical findings and theoretical views is that attempts to train a person on one task and thus bring about improvements in tasks other than the trained task are likely to fail unless the tasks are similar in terms of their elements or components.

Other scholarly views, however, allow for greater possibilities when it comes to transfer of training. Some research suggests that important moderators influence the degree of transfer resulting from training. One moderator is the degree of variability encountered during training, such that more variable training leads to greater transfer.[8, 9] Transfer may also be more likely to occur when performance of the practiced task and the transfer task depend on overlapping neural circuits. For example, some researchers observed transfer between a trained and an untrained task to the extent that both tasks activated a region of the brain called the striatum, while they observed no transfer when this region was not activated during an unpracticed task.[10] Finally, certain types of training may sharpen abilities that are so fundamental to a wide variety of tasks that performance of additional untrained tasks improves. For example, the performance of all tasks requires some degree of learning. If cognitive training helps individuals make better use of statistical/probabilistic information within a task, it could account for superior performance across a variety of untrained tasks. (For discussion of the "learning to learn" hypothesis of transfer, see endnote 11.)

While many theoretical accounts of learning reflect skepticism regarding the ability of cognitive training to improve the performance of untrained tasks, under certain conditions and with certain types of training, these effects may be observable. These theoretical accounts make it clear that it is not safe to assume that all types of cognitive training will produce meaningful benefits affecting important everyday tasks.

Empirical evidence that certain software packages and digital games are capable of improving perceptual and cognitive abilities that transfer to untrained tasks is mixed. Some studies had positive results, while others did not. And even in studies with positive results, interpretations of transfer effects aren't always straightforward. This is still a very active area of research.

Popular Approaches

There are three popular approaches to improving cognition: brain-training programs, working-memory training, and video-game training. The

ACTIVE (Advanced Cognitive Training for Independent and Vital Elderly) clinical trial was the largest test of whether brain training can improve perceptual and cognitive abilities in older adults.[12] Over 2,800 participants were randomly assigned to one of four conditions: memory training, reasoning training, speed-of-processing training, or a no-contact control group. Intervention groups received 10 training sessions, each approximately 60 to 75 minutes long (some participants also received a few booster sessions in the years following training). Transfer tasks included laboratory-based tests of cognition (proximal outcomes) and self-reported and simulated performance-based measures of daily functioning (primary outcomes).

Immediately after training, researchers observed large improvements that were specific to each type of training intervention (for example, speed-of-processing trained participants improved on laboratory tasks measuring speed but not on tasks measuring memory or reasoning). Participants maintained most of these improvements even when they were tested 10 years later. However, researchers observed no improvements on measures of everyday functioning immediately after training, one year after training, or two years after training. However, tests of participants 5 and 10 years later indicated more promise. Compared to the no-contact control group, five years after training, the reasoning group self-reported fewer daily-living problems, the speed-of-processing group was less likely to cease driving, the speed-of-processing and reasoning groups were involved in fewer at-fault automobile crashes, and the speed-of-processing group reported less of a decline in health-related quality of life. Researchers attributed these delayed effects to the facts that (1) at the start of the intervention participants were cognitively healthy, and (2) a certain amount of decline was necessary in order to reveal transfer effects.

Other studies have had less encouraging results, however. For example, Adrian Owen and colleagues randomly assigned more than 11,000 online participants between the ages of 18 and 60 to receive six weeks of reasoning training, to receive six weeks of visuospatial/attention training, or to be part of an active control group that answered trivia questions.[13] Despite the tremendous sample size, neither training group demonstrated improved general ability on a battery of neuropsychological tests.

In general, it is hard to draw straightforward conclusions from the current body of literature on brain training, even when significant effects are observed. With the notable exception of the ACTIVE trial, these studies generally focus on outcome measures based on abstract neuropsychological tests and utilize weak control groups. A recent test of a popular commercial cognitive-training program, for example, assessed transfer with an abstract digit/tone categorization task.[14] While researchers observed some evidence of transfer to neuropsychological tests of alertness and distraction, the extent to which transfer to the performance of *important everyday* tasks was unclear. As with any intervention, brain-training studies need to prove convincingly that transfer-task improvement cannot be accounted for by a placebo effect.[15] That is, researchers need to rule out any possibility that the group receiving brain training didn't improve more than the control group did simply because their treatment caused them to *expect* this outcome.[16] Julia Mayas et al. compared a group that received intense brain training to a control group that participated in discussion groups.[14] It is unclear whether participants who merely discussed issues related to aging would expect as much improvement on the transfer task compared to participants who received challenging and adaptive cognitive training.

Much of the recent cognitive-training literature has focused on the potential of working-memory training to improve IQ and, specifically, fluid intelligence (the ability to reason and to solve novel problems). Susanne Jaeggi, Martin Buschkuehl, John Jonides, and Walter Perrig first reported that training that involved juggling multiple pieces of information in the mind affected fluid intelligence.[17] Training was adaptive, as participants had to remember visual and auditory information on each trial and compare this information to the information heard and seen one, two, three, or *N* trials back (referred to as an N-back task). When participants were able to remember information more successfully, they were given more information to remember (N was increased throughout training). Compared to participants who did not receive training, participants who received N-back training improved more on transfer assessments that included problems from standard measures of IQ.

Jaeggi and colleagues interpreted the adaptive nature of their training and the necessity of working memory to solve complex problems as being supportive of transfer to measures of intelligence. However, after this initial positive finding, other scholars raised a variety of methodological criticisms of this and other working-memory-training studies.[18] Furthermore, other studies could not replicate the effect of working-memory training on fluid intelligence.[19–21] A recent meta-analysis found that when most existing studies were considered together, working-memory training appeared to have a small but reliable effect on measures of IQ.[22] But the most rigorous studies—those that included an active control group to help address the problem of placebo effects—found almost no effect at all. Given the mixed state of the literature, two problematic possibilities exist: (1) working-memory training may not improve fluid IQ, or if it has an effect, the effect may be small, and (2) important but unknown moderators may determine who benefits from this type of training and who does not.[23]

Over the past decade, some commercial and custom video games have also generated excitement about their potential to improve a variety of perceptual and cognitive abilities. This excitement has been heavily influenced by the groundbreaking work of C. Shawn Green and Daphne Bavelier.[24] Their initial study, which focused on the effects of fast-paced action video games (typically involving violent, first-person shooters), found not only that action gamers demonstrated superior visual and attentional abilities compared to nongamers, but also that nongamers could improve these abilities with just a small amount of action-game training. This finding led to dozens of additional investigations into other abilities that might be improved through action-game training. Researchers have linked superior attention, vision, processing speed, dual-tasking ability, and decision-making to action-game play through cross-sectional studies comparing gamers to nongamers, intervention studies training nongamers to play action games, or both.[25] Other studies have suggested that game training could ameliorate age-related cognitive decline.[26] Unlike focused N-back training, video games tap a variety of perceptual, cognitive, and motor processes, likely ensuring a greater degree of cognitive and neural overlap between trained and untrained tasks. This might explain the broad degree of transfer that seems to come from game training.

While this line of research is exciting, and it appears to indicate transfer that is much broader than that caused by any type of intervention investigated thus far, we must consider some important caveats when proposing to improve general cognitive abilities with video-game interventions. First, game effects do not always replicate,[27,28] again suggesting either smaller effects on cognition than previously reported or the existence of moderators that determine whether an individual might benefit from game training. Second, scholars have raised a variety of methodological criticisms of the studies that provide evidence in support of game effects.[15,29,30] Finally, as with previous types of interventions discussed here, there is a dearth of studies linking video-game interventions to better performance of meaningful everyday tasks and meaningful activities such as avoiding crashes while driving, succeeding academically or professionally, and making complex life decisions such as those involved in the purchase of a new home.

Promises, Promises

What do the sellers of cognitive-training products promise? Should consumers purchase and use them? A careful inspection reveals that most commercial brain-training companies are relatively conservative with respect to their advertised claims, at least when explicitly discussing potential improvements on everyday tasks. It is exceedingly unlikely for a company to claim that its product could help a driver avoid a dangerous crash, a worker advance his or her career, or an older adult live independently longer. Instead, claims in these commercials and advertisements are vague. They highlight improvements to more abstract qualities, such as reaction time, attention, and memory. Few specify the exact nature of these improvements—for example, reaction to what? Memories of what? These vague claims are justified in that the products' training tasks involve these abilities, and performance on the training tasks improves with practice.

The critical question, however, is the degree to which these improvements transfer to more meaningful activities. Cognitive-training advertisements typically ignore this issue. These ads typically feature product users (or actors portraying users) discussing *why* they are using the product (for example, "to remember names of people I meet," "to get ahead at work"). The companies' websites also tend to feature user anecdotes, as well as a section explaining the science behind their product and referencing completed, peer-reviewed (but sometimes non-peer-reviewed) studies. In many instances, however, these studies examine something other than the program being advertised; they assess benefits with abstract laboratory tasks rather than everyday ones; and they lack critical control conditions necessary to link improvements to the product. While pharmaceutical advertisements are strictly regulated, this is not the case for brain-fitness program advertisements. This may partly be due to the companies' lack of explicit claims regarding improvements to everyday, meaningful activities, as well as the lack of claims that their products are intended to treat specific conditions, such as age-related brain diseases.

To Be Determined

Before confidently recommending the use of brain-training programs to improve cognition meaningfully and to address age-related cognitive decline, researchers must address the following questions and issues:

- **Comparative effectiveness.** If brain-training programs and video games are in fact effective, researchers must determine the programs' comparative effectiveness. Per hour invested, how do brain-training programs and games compare to one another with respect to their ability to improve cognition meaningfully? How do they compare to other cognitively beneficial (and potentially more enjoyable) activities such as aerobic exercise,[31] digital photography, quilting, and volunteer work?[32–34] Do certain activities transfer especially well to tasks such as driving, while other activities improve the memory functions that support medication adherence? Answers to these questions would help shape recommendations regarding the amount and type of brain-training activities a given individual should engage in.
- **Intervention adherence.** As with physical exercise and pharmaceutical treatments, brain-training programs yield little to no benefit unless

people adhere to them. A recent study found that digital game-based training associated with a variety of perceptual and cognitive improvements resulted in no benefit in a sample of older adults, likely due to the fact that adherence was poor for the intervention expected to produce the largest effect.[35] The challenges of ensuring cognitive-intervention adherence may be most analogous to the challenges of promoting adherence to hypertension treatments. Given that hypertension is typically asymptomatic, treatment benefits are not readily apparent, such that the costs (e.g., drug side effects) become more salient than the real but unseen benefits. Similarly, cognitive training may not result in immediate, perceptible benefits, but it might reduce cognitive problems years in the future.[36] Thus it is important that researchers examine individual differences that predict adherence (for such an attempt with exercise, see endnote 37) as they determine how to promote adherence to brain-training games and programs.

- **Moderating variables.** Currently we know little about who benefits most from brain training. However, researchers have begun to use data from the ACTIVE trial in an attempt to answer this question. George W. Rebok et al. found that memory-training benefits were greater for participants with higher levels of education and better self-reported health. The discovery of moderating variables may help health-care professionals prescribe either general cognitive training or specific types of cognitive training. However, answering these questions will require fairly large samples to tease out the cognitive, environmental, disease, and genetic factors that make an individual more or less susceptible to the benefits of cognitive training.

- **Methodological rigor and replication.** Scholars have leveled a variety of criticisms against studies that report evidence of transfer of training from video games and brain-training programs to other tasks. These criticisms should be addressed before practitioners make strong recommendations that individuals engage in these activities. In addressing the potential of placebo effects, expectations for improvement on transfer tasks can be assessed upon completion of the intervention,[16] or in a separate group of individuals.[39] When expectations for improvement are equal for intervention and control groups, but actual transfer effects differ, placebo effects are unlikely. In addition to addressing methodological concerns, researchers should also note that the brain-training literature contains few direct replications. This is understandable because these types of studies are difficult and expensive to run. However, replication studies would be of tremendous value in answering the question of whether reliable transfer gains can be expected to result from any specific type of training. These types of studies should be incentivized.

Meaningful Measures, Outcomes, and Questions

It's a no-brainer that individuals purchase and engage in brain-training programs because they wish to perform better on certain tasks that are meaningful to them. Yet the majority of studies in the literature use relatively simple, process-pure laboratory tasks to assess transfer of benefit.[40] Few studies assess performance on simulated everyday tasks (for example, through a driving simulation), and far fewer assess real-world outcomes (e.g., automobile crash rate, loss of independence, or loss of wealth due to fraud). These types of important and meaningful outcomes can be assessed only in large-scale longitudinal studies that follow cognitively trained individuals over a decade or more.

Other important questions relate to when cognitive training should begin, how much an individual should train, and how long training gains might last. If brain training is judged to be effective, should it begin when someone is in his 20s? In her 60s? Should individuals train every day? Most days of the week? Is it better to engage in long, spaced-out training sessions or fewer, shorter training sessions? Does an individual need to continue training in order to maintain gains, or do training gains persist long after training has ceased? Is there a point of diminishing returns at which the training task becomes so automated that it no longer exercises the abilities it was designed to improve? Can people enhance the potential benefits of cognitive training if they pair it with physical activity and/or social interaction? Only a handful, if any, studies have addressed these important issues.

These are only some of the unanswered questions regarding brain training. What, if anything, can today's doctors recommend to those who wish to enhance (or to maintain) their cognitive abilities? At this point, any blanket endorsement of a certain brain-training program would be premature. Yet there appears to be enough accumulated evidence that being cognitively inactive is not a good strategy for maintaining cognitive health. Doing something to remain active and engaged is likely an investment worth making. Cognitive activity takes many forms, and there is currently little evidence suggesting that any particular software package is best at improving cognition, or that any brain-training product is better than other engaging activities, such as learning a new language or instrument, creative writing, or learning to dance.

These latter alternatives have the advantages of being inexpensive, being especially enjoyable, and providing a useful and valuable skill—even if there were no general cognitive benefits associated with them. Aerobic exercise may be one of the safest bets for those wishing to improve their cognition, as animal models and human cross-sectional and intervention studies all indicate benefits to brain function, structure, and cognition.[41] This option may be particularly beneficial because it also comes with a host of physical health benefits. Exercise would be a worthwhile investment even if it had no effect on cognition.

In the future, more precise recommendations will be possible as more evidence accumulates and the methodological rigor of intervention studies continues to advance. Large-sample studies that include real or simulated performance on important everyday tasks, extended post-training testing and observation periods (similar to those used in the ACTIVE study), and large individual-difference batteries (cognitive, genetic, neurophysiological) that assess moderators of transfer effects will be especially valuable in informing these recommendations.

12

You Say You Want a Revolution?

By Wise Young, M.D., Ph.D., and Patricia Morton, Ph.D.

Wise Young, M.D., Ph.D. came to Rutgers University in 1997 and is founding director of the W. M. Keck Center for Collaborative Neuroscience, the Richard H. Shindell Chair in Neuroscience, and a distinguished professor. In 2006, he organized a 24-center clinical trial network in China and ran trials to assess promising therapies for chronic complete spinal cord injuries. Phase II trials using umbilical cord blood mononuclear cells injected into the spinal cord, lithium, and intense locomotor therapy show promising results. He plans Phase III trials in China, India, Norway, and the U.S. Young established methylprednisolone as the first treatment for spinal cord injury, developed the first standardized rat spinal cord injury model, and founded the *Journal of Neurotrauma* and the National and International Neurotrauma Societies. Young received his B.A. from Reed College, his Ph.D. from the University of Iowa, and his M.D. from Stanford University. After a surgery internship at New York University and Bellevue Medical Center, he joined the neurosurgery department at NYU, where he became director of neurosurgery research in 1984.

Patricia Morton, Ph.D., director of planning and development for the W. M. Keck Center for Collaborative Neuroscience and an assistant professor at Rutgers University, assists in developing the Keck Center and Spinal Cord Injury Project. She organized the first New Jersey Neuroscience Symposium, the inaugural symposium for the New Jersey Commission on Spinal Cord Research, and was a founding member of the commission. She coordinated the state-by-state advocacy group, *Quest for the Cure,* and leads a seminar on spinal cord injuries, stem cells, and clinical trials: "Pushing the Frontiers, Raising the Ethical Questions."

Editor's Note: From their roles directing the W.M. Keck Center for Collaborative Neuroscience at Rutgers University, Wise Young and Patricia Morton have been on the front lines of spinal- cord- injury research for most of their careers. In this article they lean on lessons from the past, their own experience, and events still unfolding as they raise questions about the future of all scientific research.

WE ARE IN THE MIDST OF A SCIENTIFIC revolution, changing from the long-established practice of teaching patients to live with disability to the new field of regenerative medicine that utilizes stem cells and other approaches to regenerate tissues and restore function. As with all revolutions, regenerative medicine is encountering opposition on many fronts and for many reasons. Some objections are moral, others scientific. Some people object to a particular approach because they think it is scientifically unsound, while others have a vested interest in a different methodology or a favored cell or mechanism. The lack of sufficient funding will fuel these attacks because projects and careers are at stake.

This revolution raises new questions and requires new strategies. For example, scientists are trying to figure out which lessons from the past will inform the future. In their 2013 book *Decisive: How to Make Better Choices in Life and Work*, Chip and Dan Heath identify several ways that people get trapped when making decisions: "Research in psychology has revealed that our decisions are disrupted by an array of biases and irrationalities: We're overconfident. We seek out information that supports us and downplay information that doesn't. We get distracted by short-term emotions. When it comes to making choices, it seems, our brains are flawed instruments. Unfortunately, merely being aware of these shortcomings doesn't fix the problem, any more than knowing that we are nearsighted helps us to see. The real question is: How can we do better?"[1]

One of the shortcomings that Chip and Dan Heath stress is the human tendency to get trapped in binary (either/or) thinking. This raises a query: What are some ways in which moving beyond binary thinking could change the future of science? Let us suggest four areas to consider.

Collaboration *and* Competition

It's the lone researchers who generally explore scientific frontiers, but groups of people with various areas of expertise come together to consolidate advances. As new knowledge arises, these groups solidify standards that provide the platform for the next frontiers.

The battles between Jonas Salk and Albert Sabin to find a vaccine for poliomyelitis are well documented. Each scientist was driven and fervently believed in his approach. Funding from the National Foundation for Infantile Paralysis fueled the competition, and secrecy hid fundamental errors that openness might have quickly revealed. The competition strengthened each researcher's commitment to his approach, but it perhaps caused a significant delay in finding a vaccine when both Salk and Sabin initially ignored the findings of Dr. Dorothy Horstmann, the woman who discovered the actual path of the viral entry.[2]

On the other hand, an effective model of collaboration functioned between 1993 and 1996, when the National Institutes of Health (NIH) funded work on the Multicenter Animal Spinal Cord Injury Study (MASCIS). Eight leading spinal- cord injury laboratories in the United States worked together to develop and validate the first standardized rat model of spinal- cord injury. In addition to developing the model, MASCIS scientists developed outcome measures such as the Basso, Beattie, Bresnahan (BBB) locomotor scale and white-matter sparing, both of which became standards in the field. The first project of its kind funded by the NIH, MASCIS proved that people in different laboratories can work together to develop approaches and to standardize procedures, thus enhancing the work of each of the individual researchers.

The current model of individual principal investigators competing with each other for funding, and the consequent lack of collaboration, not only is expensive and inefficient but also forces scientists into undesirable binary thinking that invariably accompanies competitive endeavors. What your competitor is doing automatically becomes off-limits. A much better approach to developing new scientific discoveries would be to offer increased funding of projects that both enhance collaboration and stimulate individual initiative.

Biology *and* Technology

Biotechnology has become the buzzword of the 21st century; an entire industry has arisen around the term. But biology and technology don't always work well together. One example is the dichotomy between the electrical stimulation and cellular transplantation approaches to restoring function to people who are paralyzed.

Implemented by engineers who think in terms of electrical current, voltage, and resistance, a whole field has been built around functional electrical stimulation (FES)—using computers to deliver electrical signals to muscles and treating human muscles like robotic components. The film *RoboCop* exemplifies the thinking and limitations of this approach. In contrast, cellular transplantation is the brainchild of biologists, who think in terms of synapses, neurotransmission, and metabolism. As with FES, an entire industry has grown up around the concept of cell transplantation. Fueled at the end of the 20th century by the discovery of stem cells, this field has been dominated by paranoia and fanciful thinking illustrated in the film *Star Wars: The Clone Wars* (2008).

Biology and technology must work together in practical and realistic ways to restore meaningful function based on the best available technology and understanding of biology. A recent example provides a clear illustration of a fruitful marriage between biology and technology. Each field provides a best-of-class solution to the problem of restoring function after spinal- cord injury.

Many researchers have shown that the spinal cord can regenerate. For example, Lu et al demonstrated that rivers of axons can grow across the transection site of a rat spinal cord if you implant mesenchymal cells to form a bridge, inject the response-element binding protein cAMP to motivate neurons to grow the axons, and induce the neurons on the other side of the gap to express growth factors and "come hither" signals.[3] Likewise, Liu et al showed that silencing a single gene called PTEN can stimulate rivers of axons to grow across the injury sites in mouse spinal cords.[4]

The problem was that the rats and mice did not recover motor function even though thousands of axons grew across the injury site and made connections with neurons above and below it. It was fairly obvious why function did not recover. The regenerated axons were new, and they probably were not connecting to the neurons in the same way the old axons had. Thus, the brain had no idea which "buttons" to push to move specific muscles or how to interpret incoming signals.

Much evidence suggests that intensive locomotor training is needed in order to restore function. In fact, the most successful mobility training programs for those with spinal- cord injury are those that involve prolonged repetitive activation of desired movements as often as six hours a day, six days a week, for six months or more. Such intensive training is not only expensive but also unavailable to the majority of those who would need it.

Why is such intensive training necessary for functional recovery? It turns out that learning requires repetitive activation of synaptic connections. In Donald Hebb's[5] book, *The Organization of Behavior: A Neuropsychological Theory* (1949), he proposed that timing of synaptic activation is responsible for learning. Specifically, Hebb said, "When an axon of cell A is near enough to excite cell B and repeatedly or persistently takes part in firing it, some growth process or metabolic change takes place in one or both cells such that A's efficiency, as one of the cells firing B, is increased." Sometimes rephrased as "Neurons that fire together, wire together," the Hebbian principle has become a leading theory of neuronal learning. The formation and consolidation of synapses or connections requires synchronized activity from sensory and central sources. Intensive exercise can achieve such synaptic consolidation, while desynchronized activation weakens it.

Electrical stimulation is one way to induce synchronized activity for spinal- cord injury. Today most people carry in their pocket or purse more computer power than what once filled an entire room of mainframes. Brain-to-machine interface has demonstrated the ability to control and deliver electrical stimulation synchronized to desired activities and thereby increase synaptic consolidation and learning of regenerated fibers. Therefore, a combination of cell transplantation and electrical stimulation is the best way to restore function when neither can do it alone.

Regenerated axons are not a repaired nervous system but a new one in which new neuron-to-neuron and neuron-to-muscle connections must be learned. Initial studies show that transplanting umbilical-cord-blood stem cells in combination with intense physical therapy restores walking in people with spinal- cord injury. But the cost would be prohibitive for large numbers of people to participate in months-long walking programs. However, what if people could receive cell transplants and walk two hours a day in addition to undergoing electrical stimulation that allows them to continue to "walk" while they sleep? Injected stimulators such as the rice-size BION raise this possibility.

Hope *and* Realistic Expectations

The story of stem- cell advocacy is relevant to the balance between hope and realism. Celebrating the passage of stem- cell legislation in New Jersey in 2003, Commissioner of Health Fred M. Jacobs, M.D., J.D., proclaimed that stem-cell medicine was the most significant paradigm shift in the 40 years he had been practicing medicine. Newspapers heralded the advance, and community advocates believed that with the exception of restrictions imposed by President George W. Bush, the way was clear for miraculous cures for devastating diseases.[6]

In a movement called Quest for the Cure, impatient activists worked together to pursue stem-cell legislation at the state level. In December 2003 New Jersey passed S1909/A2840, and California's Proposition 71 followed in November 2004. Each of these bills, and others that followed, provided avenues for funding stem- cell research within its respective state. During a subsequent backlash, other states increased restrictions on—or entirely prohibited—fetal and embryonic stem- cell research.

On March 9, 2009, excitement filled the room when, in front of many disabled and ill people and their families, newly elected president Barack Obama signed an executive order lifting restrictions related to human embryonic stem cells. The president proclaimed, "Today, with the Executive Order I am about to sign, we will bring the change that so many scientists and researchers; doctors and innovators; patients and loved ones have hoped for, and fought for, these past eight years: we will lift the ban on federal funding for promising embryonic stem cell research. We will vigorously support scientists who pursue this research." Later, when the order's official guidelines were released, scientists were disappointed to discover that the new policy was more restrictive and onerous than the previous one. Even more disappointing was the fact that no increased funding was to follow.

Two years earlier, Shinya Yamanaka's discovery that skin cells can be reprogrammed to become embryonic stem cells called induced pluripotent stem (iPS) cells had changed the field and had made it unnecessary to harvest fertilized eggs in order to obtain embryonic stem cells. To date, no iPS cell has been tried in humans because of fears that these cells can produce tumors. However, the discovery established the principle that pluripotency is genetically programmed. Further recognition of the discovery's significance came when Yamanaka received the Nobel Prize in 2012 for Physiology or Medicine.

The suppression of embryonic -stem- cell research *did* lead to a renaissance of studies of adult stem cells. This soon resulted in over 4,500 clinical trials involving stem cells in the United States. The field of stem -cell therapies has turned 180 degrees from embryonic stem cells because pluripotency is not as desirable as originally thought. In fact, most scientists and clinicians consider pluripotency to be a dangerous property that must be eliminated before cells can be transplanted. For example, Geron Corporation obtained permission from the Food and Drug Administration (FDA) to transplant embryonic stem cells into people with spinal- cord injury, but only proving that less than one in a billion transplanted cells were pluripotent.

Pluripotency—the ability of make many kinds of cells—is a dangerous property if not controlled. Stem cells that find their way into the spinal cord to make a hair or a toenail or to penetrate a tissue—but end up causing a tumor—are harmful to the point where the human body has evolved to suppress pluripotency. An adult stem cell can behave like a stem cell only if it finds a niche of cells that tell the stem cell exactly what to do, which is why it is so difficult to grow stem cells in culture. Nature has developed ways to reduce the dangers of stem cells by forcing them to produce the correct type and number of cells in response to tissue requirements.

Hype comes from ignorance. When embryonic stem cells first were discovered and grown in culture, we did not understand stem cells and the implications of pluripotency. Scientists were excited about the possibility of growing any and all sorts of tissues from embryonic stem cells. The public regarded embryonic stem cells as a panacea. Religious conservatives believed that allowing embryonic stem cell research to proceed would lead to the practice of killing fetuses to treat adults. With more knowledge and greater understanding of the biology of stem cells, we now have a more balanced approach to stem cell therapies.

Real hope comes from knowledge and understanding. For example, we know from animal studies—and hopefully soon from clinical trials—that the spinal cord can regenerate. At the same time, our expectations are tempered by the observations that animals do not recover function after regeneration and that intensive exercise and training are needed to restore function. But lest overhyping hope sow the seeds of its own destruction, hope must be coupled with honest realism.[7] In the absence of understanding, scientists would do well to under promise and over deliver.

Compassion *and* Caution

Each potential treatment raises the question: When is "too soon," and when does an overly conservative approach perpetuate human suffering? Are large animal studies always required before people can be studied, and to what extent are double-blind randomized trials de rigueur? What data are sufficient to move forward with clinical trials? When is a situation so critical that immediate action is essential?

The present Ebola virus disease (EVD) epidemic has brought this debate into focus. As deaths mounted, ZMapp (Mapp Biopharmaceutical, Inc., San Diego), a potential treatment untested in clinical trials, was given to six patients. Three lived, and three died. According to the World Health Organization (September 2014 Fact Sheet N103), the average fatality rate for untreated EVD is 50 percent, with a range of 25 to 90 percent. So, what, if anything, was learned? Did a sense of urgency overcome scientific rigor? Should it have? Now one idea is that the blood of EVD survivors might impart immunity to patients, and transfusions are being given to patients without meticulous study.

The question of timing raises the important issue of whether there should be different standards for situations like the AIDS epidemic of the 1980s and the present EVD crisis—in which death is a highly probable outcome—or for conditions like spinal- cord injury, in which people are paralyzed but stable. Should victims of the former have fast-track access to untested potential treatments while victims of the latter are made to wait through a full clinical-trial process? Should there be different criteria for children as the enterovirus D68 (EV-D68) respiratory illness spreads across the United States?

One danger arising from delay in committing resources and manpower is the burgeoning industry of false promises. Whenever new possibilities arise, such as stem cells and clinical trials, so-called clinics spring up like mushrooms and offer these treatments—or *claim* to offer them—without waiting for trial results. For instance, the black market for the blood of EVD survivors was flourishing almost before the newsprint was dry. Unscrupulous opportunists lie in wait to take advantage of the desperation of the afflicted and their families.

Under what circumstances should compassion supersede caution? Should the definition of "compassionate use" be expanded to offer people the option to obtain treatments approved by the FDA for Phase III trials but at non-trial sites by companies in exchange for cost of the therapy? It behooves the scientific community to develop guidelines to balance the urgency of impending death with the dreams of those who would be healed.

Another principle advanced by the Heaths in their book *Decisive* is that of multitracking options—that is, keeping all options on the table. Politics, fear, and mistrust raise the "slippery-slope" argument, which too often results in the closure of pathways that may be beneficial. For example, would we have safe nuclear power today if fear had shut down nuclear research? Would adult stem cells have been discovered or well understood if all embryonic- stem- cell research had been terminated, as some people wanted?

There will always be tension between mechanical and biological approaches, between hope and hype, and between caution and risk—the struggle surrounding whether to test something new as early as possible despite the inherent danger or to wait too long at the potential expense of human lives. Debates about funding short-term immediate therapies or long-term potential breakthroughs will continue to rage. Are we open to radically different changes, such as shifting from developing each vaccine from scratch to the "chassis approach," used by VaxCelerate, which cut time and money for a Lassa fever vaccine from several years and billions of dollars to four months and less than a million? Critics will continue to question practices such as spending money and using manpower on heart transplants from pigs and baboons; isolating stem cells from urine; 3-D "printing" of organs; DARPA's ElectRx program, which may give humans self-healing powers[8]; and quality of life issues like that raised by Ezekiel Emanuel in his recent article "Why I Hope to Die at 75"[9]: How long is too long to live?

As we look to the future, we can learn from past revolutions. Experience enables us to anticipate. During the battle over polio vaccines, Dr. Thomas Rivers reminded the researchers, "Nothing is sacred in science; you give up the old when you find something new that is better." When we fail to follow promising leads, we freeze ourselves to the obsolete, shut-out the critically important, break our foundational commitment that science exists to help people. Every path will be bumpy, and many roads will be dead ends. There always will be scientific and moral questions without easy answers.

But the greatest ethical travesty would be to stop the science.

BOOK REVIEWS

13

Truth, Justice, and the NFL Way

Review: League of Denial: The NFL, Concussions, and the Battle for Truth

By Mark Fainaru-Wada and Steve Fainaru
Reviewed by Philip E. Stieg, M.D., Ph.D.

Philip Stieg, M.D., Ph.D., is neurosurgeon-in-chief of New York-Presbyterian/Weill Cornell Medical Center and chair and founder of the Weill Cornell Brain and Spine Center. Stieg, a neurosurgeon with expertise in cerebrovascular disorders and skull-base surgery, is also a past chair of the Congress of Neurological Surgeons, former president of the Society of University Neurosurgeons, and a widely published author and internationally known lecturer. Stieg is also one of the editors of the textbook *Intracranial Arteriovenous Malformations* (Nurse Education Series, Brigham and Women's Hospital, Boston, Massachusetts: 1989).

Our reviewer, Philip E. Stieg, a neuro-trauma consultant on the sidelines of NFL games, is no stranger to the violence of football. In his review of League of Denial *by Mark Fainaru-Wada and Steve Fainaru, Stieg finds the sports-concussion crisis to be a difficult subject. Beyond the heart-wrenching stories of the players that the authors use to illustrate the impact of chronic traumatic encephalopathy (CTE), as well as the moral and legal pressure and competition to advance the science, one undeniable fact remains: We still have much to learn about CTE and its impact on the future of professional football.*

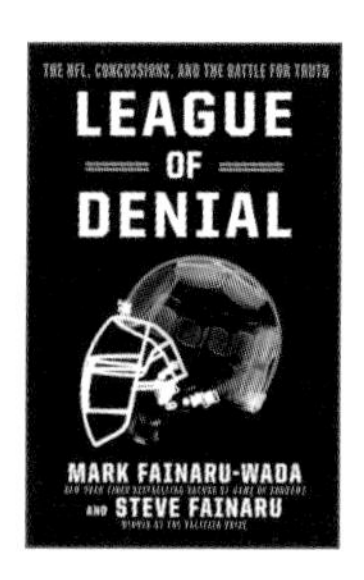

THIS ENGAGING BOOK, by respected investigative journalists (and brothers), explores the evolving story of chronic traumatic encephalopathy (CTE) among professional football players over the past four decades. It begins in 1974 at the start of the career of Mike Webster, the All-Pro football center who played 15 seasons for the Pittsburgh Steelers. Webster's death in 2002 was attributed to CTE, making him the first professional football player to be diagnosed posthumously with the brain disease. The authors use Webster and other star players—such as Dave Duerson and Junior Seau, both of whom committed suicide and were found on autopsy to have suffered from CTE—to tell the larger story of National Football League (NFL) players and league leadership as they grappled with the unfolding story of brain trauma on the gridiron.

Both the incidence of CTE and its causal relationship with concussions are controversial issues, and the science behind them is still unfolding. This book is a retrospective review of the subject, with the advantage of hindsight and the accompanying whiff of Monday-morning quarterbacking. The authors do an excellent job documenting the debate surrounding the diagnosis of CTE, along with its cause, prevalence, and outcome, and skillfully weave the personal player accounts into the more complex story of public relations and legal liability. The legal outcome remains unsettled after a federal judge in January denied the preapproval of a $765 million settlement of NFL concussion claims to cover 20,000 retired players for 65 years. The medical issues regarding CTE and the factors leading up to it, however, persist.

An open discussion about the role that repeated concussions play in the development of CTE started in 1996. The authors compare the NFL's role in that conversation to the role played earlier by tobacco companies as they refused to acknowledge any relationship between lung cancer and cigarette smoking. That comparison seems unfairly harsh, since the CTE story includes a complex interplay between willing partners and publicity seekers (players, owners, scientists, doctors) who unfortunately had to discuss a medical issue that had no clear definitions or vocabulary at the time.

Only recently have scientists agreed upon the clinical definitions of concussion and mild traumatic brain injury. Even today there is no gold-standard diagnostic test for concussion and no clear data on when it is safe for a player to return to action. For most of the time period this book covers, the science was even more speculative. Nevertheless, the authors' apparent agenda means that anyone associated with the NFL is described as self-interested and corrupted by dollars, while dissenters are characterized as world authorities, at a time when there really weren't any. Agenda aside, the authors create a highly readable account of the developing story, and I enjoyed their prose and progression of ideas.

The authors do an excellent job capturing the personal and behavioral characteristics of all the participants in this complicated scenario. The science on concussion is complex enough, and the subject tackled by the authors is even more confusing because of the personalities involved and the jockeying for personal gain that went on throughout the era. Industry-funded research has an inherent conflict of interest, and the authors accurately characterize the egos, self-promotion, and financial and reputational gain sought by all the characters in this story. What the authors don't adequately capture or characterize, however, is how all the complex interpersonal relationships and conflicts of interest obfuscated the central issue, which presumably is player safety. The authors also downplay the athletes' own role in the story. Self-promotion, greed, and conflict of interest at multiple levels made this a much more complex issue than it needed to be.

Moreover, the authors suggest throughout the book that CTE is an epidemic, although at the time of publication there was a total of only 50 reported cases, 33 of which occurred in NFL players. The hyperbolic prose helps to lead readers to the authors' preordained conclusion.

Both the NFL and the dissenters were unfortunately guilty of not following the age-old rule of the scientific method. As a scientist, I would propose a null hypothesis and take a prospective rather than retrospective approach: eliminate all the confounding variables and prospectively follow patients with the intention of demonstrating statistical significance. Future data will always be confounded by drugs, smoking, alcohol, and other risk factors, such as genetics, family history, and ethnicity. But rather than being led to the authors' conclusion, readers would be better served by being helped to understand how complex the diagnosis and treatment of mild traumatic brain injury and its possible outcome, CTE, can be.

Overall, I found *League of Denial* to be a great weekend read. The authors are to be commended for their compilation of data and description of circumstances that, although flavored, are accurate. The book clearly demonstrates that we need more data, which should be derived from independently funded studies to eliminate accusations of bias. I totally agree that research funds should be directed through the National Institutes of Health. The NFL appears to be continuing its support for definitions, protection, and rule changes. Only if all parties remain focused on the science, and on player safety, will this issue be resolved fairly.

14

Welcome to the Machine

The Future of the Mind: The Scientific Quest to Understand, Enhance, and Empower the Mind

By Michio Kaku, Ph.D.

Reviewed by Jerome Kagan, Ph.D.

Jerome Kagan, Ph.D., emeritus professor of psychology at Harvard University, was co-director of the Harvard Mind/Brain/Behavior Interfaculty Initiative. He is a pioneer in the study of cognitive and emotional development during the first decade of life, focusing on the origins of temperament, and is the author or co-author of more than 20 books, including the classic *Galen's Prophecy: Temperament in Human Nature* (Basic Books, 1994; Westview Press, 1997).

In Jerome Kagan's review of The Future of the Mind *by physicist and futurist Michio Kaku, Kagan leans on his own experience as co-director of the Harvard Mind/Brain/Behavior Interfaculty Initiative to explore a book that imagines a world where we will have the power to record, store, and transmit signals of brain activity, and where interchangeable thoughts and self-aware robots will be part of everyday life.*

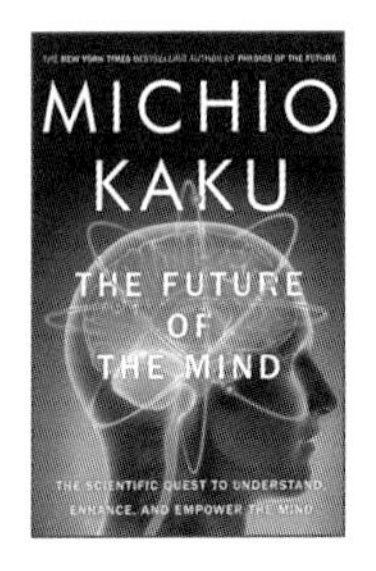

THE FIRST HUMANS PROBABLY WONDERED about the same phenomena that puzzled the 600 generations that followed: Where do the objects in the sky come from? What is the difference between living and inanimate forms? Why does like beget like? How could tiny drops of fluid become a fully formed infant? What are the origins of thoughts and feelings?

Scientists exploited appropriate technologies to provide preliminary answers to the first four questions but could not begin to determine how mental events emerge from the brain until magnetic resonance imaging (MRI), positron-emission tomography (PET), magnetoencephalography (MEG), and electroencephalographs (EEGs) became available in university laboratories. The promise of the technology's power motivated serious brooding on what has become the question sitting at the top of the stack: How does a material brain generate immaterial mental products?

Freeman Dyson, the noted theoretical physicist and mathematician, divided natural scientists into the hawks that fly above the confusing particularity of nature and the frogs that muck around in the messy details. Michio Kaku, a theoretical physicist at the City University of New York, is a high-flying hawk. The primary audience of *The Future of the Mind* is likely to assume, incorrectly, that this book will tell them how activity in neuronal collections gives rise to plans, feelings, beliefs, and actions. The book's primary mission, however, is to persuade readers of what might be possible when future machines can record the electrical and magnetic signals accompanying a person's thoughts and emotions, store them on a disc, and transfer them to a robot or to sensors in another person's brain. One day, Kaku imagines, a young bride might send a computer the brain signals she was generating on her honeymoon so that years later her grown daughter could pick up a disc with these signals and relive her mother's happy moments. This book is intended not for skeptics but for the curious nonscientist who, like Kaku, enjoys Star Trek movies and H. G. Wells's *The War of the Worlds.*

Kaku's narrative ignores the initial phases of the cascade that begins with an event generating a brain profile followed by an initial psychological outcome. These are the events many neuroscientists hope to understand. Kaku's interest, however, is in the cascade that follows the initial brain state brought on by an event or thought that leads, in turn, to a sequence of psychological outcomes and new brain states.

The first two chapters of *The Future of the Mind* contain lean summaries of the brain, the new technologies, and a discussion of consciousness, which Kaku defines as a model of the world designed to accomplish a goal. He posits three levels of consciousness, ranging from reptiles who occupy Level I to mammals at Level II and humans at the highest level. A level is defined by the number of feedback loops required for an animal to interact with other members of its group. Kaku illustrates this metrical conception by asking readers to imagine a group of 10 wolves who can display any of 15 responses when they interact with another animal. Because the product of 10 by 15 is 150, he assigns a value of Level II: 150 to each wolf's consciousness.

Kaku's simple drawings of brain circuits that are presumed to be the foundations of select psychological processes resemble Freud's hand-drawn illustration locating the ego in the prefrontal cortex, the id in the posterior cortex, and the superego in the temporal lobe. He supports his bold ideas with cherry-picked studies that ignore alternative explanations as well as failed replications. Kaku's arguments rely heavily on the interpretations of blood-flow profiles generated by functional magnetic resonance imaging (fMRI) scanners. However, he fails to tell readers that many experts remain unsure about the causes and meaning of these profiles. Nor does he acknowledge that adults watching the same film generate distinctive blood-flow patterns in the frontal lobe because of idiosyncratic interpretations of the scenes.

Kaku's prediction that the brain signals produced by one person's thoughts would generate the same thoughts in another individual whose brain received these signals is seriously flawed because the chemistry of the recipient's brain makes a critical contribution to his or her resulting psychological state. Hence, if the neurochemistry of the sender and the recipient were different, which is likely, the latter would not duplicate the mental state of the former. I am afraid no woman will be able to experience the pleasures of her mother's honeymoon.

The author's indifference to inconvenient facts that would weaken his argument reminds me of quantum physics pioneer Wolfgang Pauli's irritation with young Werner Heisenberg, who had told Pauli that he had a unified theory of matter that was missing only a few details. The next day Pauli sent several friends a single piece of paper containing a blank rectangle in the shape of a frame for a painting, along with the sentence, "This is to show that I can paint like Titian; only the technical details are missing."

The remaining 13 chapters are optimistic predictions of the consequences of an increasing power to record, store, and transmit signals of brain activity. The clearly written prose covers the transmission of memories from one person to another, robots controlled by thoughts, and transcranial magnetic stimulation (TMS) used to control brain states and thoughts. These paragraphs are interrupted, often abruptly, with brief forays reflecting Kaku's positions on dreams, genes, optogenetics, intelligence, savants, Einstein's brain, and mental illness. These sidebars, designed to provoke moments of awe, will do so in readers who do not have a deep understanding of neuroscience.

Psychiatrists will be unhappy with Kaku's claim that "mental illness is caused by the disruption of the delicate checks and balances between competing feedback loops that simulate the future." This atypical definition is accompanied by a table, which locates paranoia in disrupted loops in the amygdala and prefrontal cortex, schizophrenia in disruptions in the left temporal lobe and anterior cingulate, bipolar disease in the loops connecting the hemispheres, and obsessive-compulsive disorder in compromised feedback loops in the orbitofrontal and cingulate cortices. Surprisingly, Kaku fails to acknowledge that the best predictor of most mental illnesses, across varied societies, is spending one's childhood years with parents who are not well educated and poor.

Throughout the book, Kaku is loyal to the late British mathematician Alan Turing's suggestion that if an observer cannot tell whether an outcome—say, a particular move in a chess game—is the product of a machine or a person, it is reasonable to assume that machines and people rely on the same mechanisms. Two photographs of the Empire State Building, one taken with a Leica that used film and one with a digital camera, reveal the flaw in that premise. Most viewers could not distinguish between pictures that were the products of different mechanisms.

Such a lively narrative dealing with a domain that is foreign to the author's training has to have some errors. For instance, Kaku has Nobel Prize winner Eric Kandel working with the late German theoretical physicist Max Planck in Tübingen and asserts that area V1 of the visual cortex represents whole objects.

I suspect that before submitting the final galleys, Kaku realized, or may have been told by an editor, that he was too accepting of a deterministic materialism that made each human a slave of his or her neurons. Perhaps that is why the last sentence of the appendix contains a caveat: "In the end we are still masters of our destiny." Thank you, Professor Kaku.

15

Standing in His Shoes

The Reason I Jump: The Inner Voice of a Thirteen-Year-Old Boy with Autism

By Naoki Higashida

Reviewed by Temple Grandin, Ph.D.

Temple Grandin, Ph.D., is a professor of animal science at Colorado State University, a best-selling author, and a consultant/designer to the livestock industry and major fast-food restaurant companies on humane slaughtering systems and practices. Half the cattle in the United States are handled in equipment she has designed. She also created the "huge box," a device to calm those with autism. Two books she has authored, *Animals in Translation* and *Animals Make US Human*, were *New York Times* bestsellers. The subject of an award-winning 2010 biographical HBO film, *Temple Grandin*, she was also listed in the *Time 100* list of most influential people in the world in the "Heroes" category. Grandin earned her M.S. in animal science at Arizona State University and her Ph.D. in animal science from the University of Illinois in 1989.

In Temple Grandin's review of The Reason I Jump *by Naoki Higashida, she relates her own experience living with and studying autism to better understand the mind of a remarkable 13-year-old Japanese boy with severe autism. Structured as a series of questions that a nonautistic person might ask an autistic one, Naoki's book is translated by David Mitchell (author of the novel* Cloud Atlas*) and his wife, Keiko Yoshida, and contains an impassioned introduction in which Mitchell discusses his experience with his own autistic child.*

WHEN I RECEIVED A REVIEW COPY of *The Reason I Jump* from the publisher, I set this highly insightful book aside because both the foreword and the cover letter from the editor failed to provide a sufficient description of Naoki Higashida's ability to communicate fully and independently. When I was later asked to review this book, I was confident that Naoki, a 13-year-old nonverbal child with autism, was not using the controversial method of facilitated communication, in which a person supports the wrist of the nonverbal person with autism. When this method is used, the facilitator is often the true author. Naoki's book belongs to the other class of writings: those that come from nonverbal individuals with autism who can communicate fully independently with no wrist support.

Some of Naoki's experiences with sensory issues are similar to other accounts from nonverbal individuals, such as Tito Rajarshi Mukhopadhyay's *How Can I Talk if My Lips Don't Move?: Inside My Autistic Mind* (2008) and Arthur and Carly Fleischmann's *Carly's Voice: Breaking Through Autism* (2012). Tito, Carly, and Naoki all have a kind of "locked-in" syndrome, whereby an intelligent mind is trapped inside a body that has difficulty controlling movements. Tremendous effort is required to pay attention. I had an opportunity to visit Tito in a quiet medical library. When he arrived, he flapped and ran around. Deliberately selecting an image that was totally novel for him to comment on, I showed him a picture of an astronaut on a horse from an old *Scientific American*. He quickly typed "Apollo 11 on a horse." Tito has to work really hard to maintain attention, and he could answer only three short questions before he needed a rest. In their books, Tito and Carly describe problems with maintaining attention and screening out a bombardment of sensory stimuli around them. When I met Tito, he typed his answers very quickly and then jumped up and flapped his hands.

The title of Naoki's book is very appropriate. He explains why he needs movement in order to determine where his body is located in space: "The reason is that imitating movement is difficult for people with autism because we don't know our own body parts well." Later, he writes, "When I am jumping, I can feel my body parts really well." He goes on to write, "When I am not moving, it feels like my soul is detaching from my body." Tito's book has similar descriptions of a thinking self that is separate from the actions of his body. Movements such as spinning provide calm and bliss. Naoki writes, "Just watching spinning things fills me with everlasting bliss." He also explains how flicking his fingers in front of his eyes provides light in a pleasant, filtered manner. I can relate to this. When I was little, I spun things and dribbled sand through my hands. It was like taking a drug. If my teachers and mother had let me spin things all day, I never would have developed because my brain would have been shut off from the world. To calm down I was allowed to spin things for an hour after lunch, but at other times spinning was not allowed. A child with more severe sensory issues than mine may need more rest periods in between teaching periods.

Naoki is very clear that people with autism want to be social. He values the company of other people. In an early chapter of my book, *Thinking in Pictures: My Life with Autism* (Grandin 1995), I hypothesized that autism may be on an emotional cognitive continuum. People who are nonverbal or have great sensory problems may be more socially and emotionally "normal" than are fully verbal individuals with high-functioning autism or Asperger's syndrome. The works of Tito, Carly, and Naoki support that hypothesis. Carly had normal teenage girl interests locked in a body she had difficulty controlling.

There are some similarities in the way Naoki searches his database for memories and my visual thinking. He explains why he repeats questions: "Firing the question back is a way of sifting through our memories to pick up clues about what the question was asking. We understand the question okay but can't answer it until we fish the right memory pictures in our heads." It seems like his memory is a less organized version of how I search for visual memories. The difference is that I have great control over those associations. I can consciously "search" my visual image database by using keywords, just as Google Images does. Writes Noaki, "but in the case of people with autism, memories are not stored in a clear order." Both Naoki and I have problems with remembering long sequences. His problems are more severe than mine. Other similarities between us include the fear of certain noises and the tendency to look at the details of an object first before seeing the whole.

Research shows that autism is a disorder of the brain's white-matter connections, which provide what I call interoffice communications between the brain's parts. Based on autobiographies of both fully verbal and nonverbal people, it appears that in nonverbal people, systems that integrate sensory information are partly disconnected. A person with nonverbal autism may be working with a condition very different from that of a fully verbal "Asperger's" type who is socially awkward. On the nonverbal end of the spectrum, autism may be more of a disorder of integration of sensory and motor systems, while on the fully verbal end, autism may mean a lack of brain connections involved in social relatedness. My old continuum in my chapter "How People with Autism Think" may be right. Nonverbal individuals may have a more normal social-emotional brain locked in a body that he or she has difficulty controlling, while their visual and auditory systems provide distorted sensory input.

I wish *The Reason I Jump* included more documentation on Naoki's ability to communicate independently. It should have included descriptions of how he was taught, in either the foreword or the afterword. But the book is an important addition to autobiographical accounts from nonverbal individuals with autism. Everybody who is working with nonverbal individuals with autism should read it.

15

Funny Science

Ha! The Science of When We Laugh and *Why and The Humor Code: A Global Search for What Makes Things Funny*

By Scott Weems, Ph.D., Peter McGraw, Ph.D., and Joel Warner
Reviewed by Robert R. Provine, Ph.D.

Robert R. Provine, Ph.D., is professor of psychology and neuroscience at the University of Maryland. After training in developmental neuroscience at Washington University and investigating neurobehavioral development in many species, he developed a novel, low-tech approach to human brain mechanisms that he terms "sidewalk neuroscience," which is based on the analysis of simple instincts such as laughing and yawning. As a bonus, the contagion of these behaviors provides an entrée to the neurological basis of social behavior. Provine is a fellow of the Association for Psychological Science and the American Association for the Advancement of Science. His research is described in *Laughter: A Scientific Investigation* and *Curious Behavior: Yawning, Laughing, Hiccupping, and Beyond*.

In Robert Provine's review of Ha: The Science of When We Laugh and Why *and* The Humor Code: A Global Search for What Makes Things Funny, *he leans on his own analysis of simple instincts such as laughing and yawning and his research for his own book,* Laughter: A Scientific Investigation.

THE IMPORTANCE OF HUMOR IS SUGGESTED by the stature of those who have studied it. This formidable group includes Plato, Aristotle, Hobbes, Kant, Schopenhauer, and Darwin. The difficulty of the topic is indicated by our continuing effort to understand it. In contrast to the long history of philosophical analyses, empirically based humor study is little more than 100 years old. Two recent books for general audiences provide very different progress reports from the frontiers of humor science.

The first, *Ha: The Science of When We Laugh and Why*, by Scott Weems, a cognitive neuroscientist and postdoctoral research associate at the University of Maryland, takes us on a lighthearted tour of things funny. Mindful of the cliché that analysis kills humor, Weems starts with an amusing anecdote about groundbreaking comedian Lenny Bruce and maintains a high humor quotient throughout, mostly straddling the thin line between entertainment and revelation. Can some popularizers of humor studies try too hard to be funny? Borrowing a quote from humor scholar Victor Raskin in Weems' introduction, "[P]sychiatrists don't try to sound neurotic or delusional when describing schizophrenia, so why should humor researchers try to be funny?" What about reviewers of books about humor? It's informative that no one wants to be seen as lacking a sense of humor.

When not trying to entertain, Weems avers, ". . . humor and its most common symptom—laughter—are by-products of possessing brains which rely on conflict," and this conflict is desirable because it encourages adaptability. Further, humor is "closely associated with nearly every aspect of human cognition," "the healthiest way to stay cognitively sharp" and "strongly related to intelligence." In case you missed the point, Weems reminds us, "Nearly every aspect of our lives is improved by focusing on humor." Weems makes a good point about the variety of humor processing, given the many kinds of humor, from pun to pratfall, and its many channels of delivery, from spoken word to vision. As he suggests, humor really is a kind of IQ test—which is a theme of the delightful French film *Ridicule*.

If you want your cognitive neuroscience with a smiley face, Weems is your man. He caters to the feel-good, be-happy modern audience of positive psychology by underplaying laughter's dark side, which was a concern of the ancients. Plato, for example, was motivated by fear of laughter's power, not improving health of the quality of his stand-up. If you doubt the danger of laughter, ask a politician earning mention on a late-night comedy show, even one not named [Anthony] Weiner. Carelessly targeted laughter can trigger a beating.

Few people deny that laughter provides pleasure, but can we really "laugh our way to health," as suggested by clown/physician Patch Adams and the late writer/editor Norman Cousins (*Anatomy of an Illness as Perceived by the Patient*)? Weems is optimistic about the prospects of medicinal laughter. He underplays contrary evidence, such as the cited large-scale, long-term study of Howard Friedman at the University of California, Riverside, which indicated that conscientiousness, not humor, predicts longevity. Better news comes from pain studies that report an analgesic effect of comedy and laughter. Perhaps we are expecting too much from laughter, a vocalization that, like speech, evolved to change the behavior of other people, not to improve our health.

The neuroscience of Weems is of the "my-brain-made-me-do-it" variety, based on the notion that, when faced with a clash of ideas, the brain's conflict detector (the anterior cingulate) fires up, provides a dose of feel-good dopamine, and somehow yields ha-ha. Does this casual "neurologizing" earn its keep? The answer is a tentative maybe. Even devout worshipers at the Church of Neuroscience may question the limits of imaging to understand a joke, or of dopamine to understand its reward.

Weems may be pardoned for not solving the problem of humor, one of history's oldest and thorniest. But complex problems need not be confronted head-on, as they are here, even when bolstered by brain images. Why not adopt the simple system approach useful in attacking other complex biological systems, thereby focusing on the simple act of "ha" of the book's title instead of wading into the swamp of cognition? The simple systems approach, guided by an evolutionary perspective, has registered counterintuitive discoveries uncited by Weems. For example, the modern human ha-ha evolved from the ancestral pant-pant—the sound of labored breathing of tickle and the rough-and-tumble of our primate ancestors. Ha-ha is literally the sound of play that announces, "This is play; I'm not attacking you."

Overall, *Ha* delivers a genial, mostly derivative, neurologically oriented introduction to humor science. It easily could have been a much stronger and more versatile book. In present form, it's often difficult to identify core ideas and to follow arguments, and its thinly veiled advocacy reduces its intellectual heft. The book has a rudimentary index, no bibliography, no citations in the text, and only select references in the endnotes for each chapter, leaving unclear who did what and frustrating those who want to read further or to check facts. Topics and people cited in the text may or may not appear in the index. For example, Victor Raskin appears in the index, but not regarding his quotation used above. The potential of the book as an authoritative reference is diminished by its casual and incomplete referencing, not its breezy style.

In *The Humor Code: A Global Search for What Makes Things Funny,* authors Peter McGraw and Joel Warner team up to present a flamboyant road show in the spirit of the Bob Hope and Bing Crosby films of yore. Journalist/writer Warner is the straight man and chronicler of the antics of "Professor Pete" McGraw, University of Colorado professor of business, psychologist, humor researcher, and wannabe comedian, as they tour the world in search of the holy grail of humor. The book grew out of a series of articles that Warner published about McGraw in *Wired* and elsewhere, and the duo sometimes communicates more about the human condition than about the science of mirth, but I won't quibble. This is a delightful conception, executed with verve, warmth, and, well, humor.

With the authors at the wheel, readers join an investigation of comedy clubs in Los Angeles, *New Yorker* cartoonists in Manhattan, the site of the laughter epidemic in Tanzania (a topic in *Ha*), cultural challenges in Japan, besieged cartoonists of the prophet Muhammad in Denmark, humor in troubled Palestine, and Patch Adams's medicinal clowning in Peru. The Grand Tour climaxes at the Just for Laughs festival in Montreal, where Professor Pete put his hard-won comedic knowledge to practical test in a stand-up routine before a major-league audience. Pete didn't kill, but he got some laughs and survived with his dignity intact. Although the device of the tour is a bit contrived, it yields fresh material involving conversations with leading players and earns its keep. I wonder who paid for it.

A major theme in the book is the worldwide test of McGraw's Benign Violation Theory, which posits that in order to be funny, an event must be a violation of the norm, must seem benign, and must satisfy both conditions simultaneously. In other words, slipping on a banana peel is funny only when the victim is not injured. In marketing their idea, the authors learned that humor scholars are a tough crowd, where upstarts can earn a knee in the groin from irascible elders. Given that the book is not a theoretical treatise, I'll give McGraw a pass on his plausible premise—plenty of other humor experts will gladly provide a critique, with the enthusiasm of lions stalking a wounded wildebeest.

The Humor Code should not be criticized for what it is not: It is not a textbook of humor/laughter science, a how-to book for aspiring comedians, or a comprehensive monograph about humor theory. What it does very well is to introduce the vast, engaging, challenging terrain of the comic and to inspire readers to explore further. The thoughtful and detailed endnotes/bibliography and index will assist readers wishing to do the latter.

Peter McGraw in *The Humor Code* and Scott Weems in *Ha* both conclude their books with a display of their stand-up skills, prompting anticipation of face-to-face comedic combat between these scientists of the funny.

Note from the author: Readers wanting to learn more about humor from a carefully reasoned evolutionary, cognitive, and philosophical perspective are directed to the excellent *Inside Jokes: Using Humor to Reverse-Engineer the Mind*, by Hurley, Dennett, and Adams. My book, *Laughter: A Scientific Investigation*, complements the above with an analysis of the vocal act of laughter. Those seeking an enjoyable, wide-ranging perspective about what neuroscience can and can't tell us about how we feel will find wise counsel in Frazzetto's *Joy, Guilt, Anger, Love: What Neuroscience Can—and Can't—Tell Us About How We Feel.*

16

The Long and Winding Road

Review: Madness and Memory: The Discovery of Prions— A New Biological Principle of Disease

By Stanley B. Prusiner, M.D.

Reviewed by Guy McKhann, M.D.

Guy McKhann, M.D., the scientific advisor to the Dana Foundation, studies neurological outcomes following coronary artery bypass grafting and the elucidation of the mechanism of a form of Guillen Barre Syndrome. He has also been active in defining the criteria for Alzheimer's disease. McKhann received his B.S. degree from Harvard University and obtained his doctoral degree from Yale Medical School. After a period of time at the National Institute of Neurological Disorders and Stroke, he took his residency in pediatric neurology at Massachusetts General Hospital. His first academic position was at Stanford University, where he founded the Pediatric Neurology service. He then moved to Johns Hopkins University Medical Center, where he was the first director of the neurology department.

In his review of Madness and Memory *by Stanley B. Prusiner, M.D., Guy McKhann (scientific consultant for the Dana Foundation) leans on his own longtime relationship with the author and many of the scientists and institutions that played a role in the discovery of prions. Prions, which are infectious proteins that cause neural degeneration, are responsible for ravaging the brains of animals suffering from scrapie and mad cow disease, and of humans with a variant of mad cow disease and Creutzfeldt-Jakob disease.*

IN 1968 JAMES WATSON, a Nobel Prize winner for the discovery of the structure of DNA, surprised the scientifically oriented world with the publication of *The Double Helix: A Personal Account of the Discovery of the Structure of DNA*. Watson's book was filled with very personal observations about his colleagues, partners, and competitors alike. Originally the Harvard University Press had agreed to publish it, but after its lawyers reviewed the manuscript, the publisher reneged. Atheneum eventually stepped in, and more than one million copies were sold.

Now comes *Madness and Memory*, which has the feel of a successor book. Like Watson, the author, Stan Prusiner, is a Nobel Prize winner, who single-handedly worked out a new mechanism of disease, the infectious protein, "prion." Both books provide an inside look at how science was done, but the conditions were quite different. Watson and his colleague, Francis Crick, were determined to find the structure of a known molecule, DNA. They accomplished their goal in a short period of time—two to three years. Prusiner, on the other hand, describes in his book the steps to the discovery of a whole new mechanism of disease and its application. The process began in 1978 and continues today.

Before Prusiner began his studies, research in this area focused on two rare human neurodegenerative diseases, Creutzfeldt-Jakob Disease (CJD) and kuru, found in the Fore tribe in New Guinea. Work expanded to include a third disease, which was called scrapie because afflicted sheep scraped off much of their wool by rubbing against fences and trees, presumably because they itched. All three diseases had distinctive neuropathology but no clear-cut mechanism. What made them unique was that they were transmissible.

Prusiner's book provides the backstory behind the discovery of the prion. It starts with the dilutions of fragments from the brains of people who died of kuru or CJD. The fragments were injected into the brains of normal chimpanzees. After a couple years of incubation, the hosts developed neurological symptoms and neuropathology resembling that of the human disease. Carleton Gajdusek and his colleague, Joe Gibbs, performed the day-to-day work of these transmission studies in the late 1960s. They sometimes called the diseases they studied transmissible spongiform encephalopathies. *Spongiform* refers to the appearance of sponge-like holes in the brain. Gajdusek and Gibbs did not find the evidence of inflammation that is the hallmark of a virus infection.

Gajdusek was the guru in this field and received the Noble Prize in 1976 for these transmission studies. Prusiner's interactions with Gajdusek included proposing to join his laboratory, accompanying him to see kuru patients in New Guinea, and fighting off Gajdusek's claims that he had been first to discover the protein nature of the scrapie agent. Gajdusek was a brilliant, strange person, and Prusiner captures his unique personality well.

When Prusiner entered the field, researchers had yet to identify the agent that was being transmitted in the cases of kuru and CJD. Virologists assumed that the infectious agent was some form of virus, which they labeled an "unconventional " or "slow" virus. Work shifted to scrapie because this disease was transmissible to mice and thought not to be transmissible to humans (people did not like working with a human pathogen). Thus, another description for this line of investigation was a search for the "scrapie agent."

Following medical school and research at the National Institutes of Health (NIH), Prusiner began a neurology residency at the University of California, San Francisco (UCSF). He was not at all sure what he wanted to do. Early in his residency he evaluated a patient with CJD and followed her through the course of her illness to her death. He became hooked on studying this mysterious, fatal disease and its possible mechanisms.

Madness and Memory prominently displays the personality traits that stood Prusiner in good stead over years of research. First, he was fearless. He brushed aside or circumvented obstacles that would have sidelined mere mortals. Second, he lacked a background in virology, as well as many other aspects of biology that would have influenced his work. What Prusiner did have was a background in protein chemistry. At NIH he worked with Earl Stadtman, a noted enzymologist who ran a laboratory with the goal of purifying of enzymes. So, Prusiner approached the scrapie problem not as if he were trying to find a small virus, but as if he might find a protein. Even if he did not actually start with this approach, that's how his research evolved.

The only way to measure the amount of scrapie agent in an infected brain was to inject it into mice and wait for up to 200 days—the amount of time it might take for the mice to show signs of neurological dysfunction. This method required an astronomical number of mice. Prusiner later shifted to hamsters, which had shorter incubation times. The switch still didn't solve the numbers problem, however; Prusiner and his colleagues were using up to 1,600 hamsters at a time. One of the side issues he discusses in *Madness and Memory* is his constant search for space in which to do his research and his invention of new approaches under space constraints.

Long incubation times meant low research productivity. After three years Prusiner had only three relatively insignificant publications. He was in serious trouble when it came to securing further research support and getting tenure at UCSF. Fortunately, as he describes, senior administrators at UCSF and private foundations recognized the potential of his work and came through with academic and financial support.

It wasn't until 1982 that Prusiner's first significant, and perhaps most important, paper was published in the journal *Science*. His account of the challenges involved in getting the paper published makes for fascinating reading. The editors originally accepted the paper after some revisions but then sat on it for months, apparently because they were afraid of the potential reactions of the scientific community. In this paper Prusiner made three points that became quite controversial. One, he had applied a number of techniques to his preparations to destroy or inactivate any nucleic acids (DNA or RNA), but there was no loss of transmissibility. He was not finding any evidence of components of a virus. Two, he proposed that the particle scrapie agent might be a protein, and that a protein alone was the transmissible agent. Three, for this infectious protein he invented a name, "prion," derived from the words protein and infectious. Thereafter he used the name prion to describe the scrapie agent.

The *Science* editors' fear was justified; that paper did blow the lid off. As Prusiner writes, "It is difficult to convey the level of animosity that both the word 'prion' and the prion concept engendered." Prusiner was proposing a new mechanism of disease, and the large majority of investigators didn't like it. They were convinced that the scrapie agent was some form of virus (some scientists still are) and Prusiner just didn't know how to find it. Prusiner's subsequent presentations at meetings often degenerated into shouting matches. At one point he tried to open the door for collaboration with a British colleague, who insisted that Prusiner no longer use the word prion and that he, the British colleague, monitor any future papers that came from Prusiner's lab. That particular collaboration never materialized.

Once Prusiner was convinced that he had the protein, he went on to make some enticing, quite original discoveries. With preparations enriched for prions, he found that the prions appeared as rods in the electron microscope—presumably as aggregates of smaller versions. Again with enriched preparations, he was able to determine the amino-acid composition of a prion. He also was able to make an antibody to the prion and thus greatly enhance his ability to characterize it.

But Prusiner made a startling discovery that almost undid him: healthy brains contained a protein that appeared to be identical to the "scrapie agent," with the same amino-acid sequence. How could the infectious prion be a component of healthy brains? Prusiner entertained the possibility that the difference between an infectious prion and a noninfectious protein could be their shapes.

At the time, scientists thought that a protein's amino acid sequence dictated its shape, but Prusiner's proteins had the same sequence. Several lines of investigation yielded evidence that the conversion of the normal protein to the infectious prion involved an induced change in shape—a different folding of the protein. Thus, new terminology evolved once again:

the "normal" protein was referred to as PrPC, and the disease-causing protein became PrPSc. A protein-turned-prion could induce another protein to misfold into a prion.

Two other aspects of this saga brought the whole area of research beyond the study of rare diseases: (1) the appearance of mad cow disease, and (2) the possible role of prions in other diseases in which proteins may play a role, specifically Parkinson's disease and Alzheimer's disease.

Mad cow disease first appeared in England in 1986 and eventually affected at least 180,000 British cattle. Four million additional cattle were slaughtered in an attempt to limit the disease's spread. Many countries instituted a ban on British beef, and this ban had profound economic consequences. What was even more unnerving was that the disease was transmitted to people in at least 160 cases—and, given the disease's long incubation time, the time from exposure to the appearance of disease, health officials feared there would be many more cases. Prusiner's concept of the prion was at the forefront of the discussion; prions were now not only a mechanism of rare diseases but also an essential part of an international health problem.

The book's last three chapters discuss the current state of things. Way back in 1984, Prusiner suggested that other degenerative nervous-system diseases might have a prion mechanism. At the time, people paid little attention to this assertion. Today, however, it is front and center. Diseases such as Alzheimer's, Parkinson's, amyotrophic lateral sclerosis (ALS), and frontotemporal dementia are all dependent on specific proteins that may have the properties of a prion, suggesting a potential mechanism and new approaches to therapy. Once again, Prusiner has pioneered a new direction of thought despite the skepticism of many of his colleagues.

For people in scientific fields, Prusiner's book is a stepwise account of his remarkable achievements—some logical extensions, and others serendipity. Nonscientific readers will learn that progress is not a smooth upward curve; it involves many setbacks and periods of uncertainty. The book will enlighten and inspire you, regardless of your background.

Stanley B. Prusiner is a Dana Foundation grantee.

Endnotes

1
Your Brain Under the Microscope

1. Gurdon, J.B. and V. Uehlinger, "Fertile" intestine nuclei. Nature, 1966. 210(5042): p. 1240-1.
2. Briggs, R. and T.J. King, Transplantation of Living Nuclei From Blastula Cells into Enucleated Frogs' Eggs. Proc Natl Acad Sci U S A, 1952. 38(5): p. 455-63.
3. Wilmut, I., et al., Viable offspring derived from fetal and adult mammalian cells. Nature, 1997. 385(6619): p. 810-3.
4. Takahashi, K. and S. Yamanaka, Induction of pluripotent stem cells from mouse embryonic and adult fibroblast cultures by defined factors. Cell, 2006. 126(4): p. 663-76.
5. Takahashi, K., et al., Induction of pluripotent stem cells from adult human fibroblasts by defined factors. Cell, 2007. 131(5): p. 861-72.
6. Yu, J., et al., Induced pluripotent stem cell lines derived from human somatic cells. Science, 2007. 318(5858): p. 1917-20.
7. Wernig, M., et al., In vitro reprogramming of fibroblasts into a pluripotent ES-cell-like state. Nature, 2007. 448(7151): p. 318-24.
8. Marchetto, M.C., et al., Non-cell-autonomous effect of human SOD1 G37R astrocytes on motor neurons derived from human embryonic stem cells. Cell Stem Cell, 2008. 3(6): p. 649-57.
9. Liu, Y., et al., Directed differentiation of forebrain GABA interneurons from human pluripotent stem cells. Nat Protoc, 2013. 8(9): p. 1670-9.
10. Maroof, A.M., et al., Directed differentiation and functional maturation of cortical interneurons from human embryonic stem cells. Cell Stem Cell, 2013. 12(5): p. 559-72.
11. Brennand, K.J., et al., Modelling schizophrenia using human induced pluripotent stem cells. Nature, 2011. 473(7346): p. 221-5.
12. Israel, M.A., et al., Probing sporadic and familial Alzheimer's disease using induced pluripotent stem cells. Nature, 2012. 482(7384): p. 216-20.
13. Burkhardt, M.F., et al., A cellular model for sporadic ALS using patient-derived induced pluripotent stem cells. Mol Cell Neurosci, 2013. 56: p. 355-64.
14. Kola, I. and J. Landis, Can the pharmaceutical industry reduce attrition rates? Nat Rev Drug Discov, 2004. 3(8): p. 711-5.
15. Egawa, N., et al., Drug screening for ALS using patient-specific induced pluripotent stem cells. Sci Transl Med, 2012. 4(145): p. 145ra104.
16. Boissart, C., et al., Differentiation from human pluripotent stem cells of cortical neurons of the superficial layers amenable to psychiatric disease modeling and

high-throughput drug screening. Transl Psychiatry, 2013. 3: p. e294.
17. Peng, J., et al., Using human pluripotent stem cell-derived dopaminergic neurons to evaluate candidate Parkinson's disease therapeutic agents in MPP+ and rotenone models. J Biomol Screen, 2013. 18(5): p. 522-33.
18. Scott, C.W., M.F. Peters, and Y.P. Dragan, Human induced pluripotent stem cells and their use in drug discovery for toxicity testing. Toxicol Lett, 2013. 219(1): p. 49-58.
19. Pick, M., et al., Clone- and gene-specific aberrations of parental imprinting in human induced pluripotent stem cells. Stem Cells, 2009. 27(11): p. 2686-90.
20. Marchetto, M.C., et al., Transcriptional signature and memory retention of human-induced pluripotent stem cells. PLoS One, 2009. 4(9): p. e7076.
21. Hu, B.Y., et al., Neural differentiation of human induced pluripotent stem cells follows developmental principles but with variable potency. Proc Natl Acad Sci U S A, 2010. 107(9): p. 4335-40.
22. Ghosh, Z., et al., Persistent donor cell gene expression among human induced pluripotent stem cells contributes to differences with human embryonic stem cells. PLoS One, 2010. 5(2): p. e8975.
23. Fusaki, N., et al., Efficient induction of transgene-free human pluripotent stem cells using a vector based on Sendai virus, an RNA virus that does not integrate into the host genome. Proc Jpn Acad Ser B Phys Biol Sci, 2009. 85(8): p. 348-62.
24. Soldner, F., et al., Parkinson's disease patient-derived induced pluripotent stem cells free of viral reprogramming factors. Cell, 2009. 136(5): p. 964-77.
25. Zhou, H., et al., Generation of induced pluripotent stem cells using recombinant proteins. Cell Stem Cell, 2009. 4(5): p. 381-4.
26. Zou, J., et al., Oxidase-deficient neutrophils from X-linked chronic granulomatous disease iPS cells: functional correction by zinc finger nuclease-mediated safe harbor targeting. Blood, 2011. 117(21): p. 5561-72.
27. Miller, J.C., et al., A TALE nuclease architecture for efficient genome editing. Nat Biotechnol, 2011. 29(2): p. 143-8.
28. Cong, L., et al., Multiplex genome engineering using CRISPR/Cas systems. Science, 2013. 339(6121): p. 819-23.
29. Mali, P., et al., RNA-guided human genome engineering via Cas9. Science, 2013. 339(6121): p. 823-6.
30. Homma, K., et al., Developing rods transplanted into the degenerating retina of Crx-knockout mice exhibit neural activity similar to native photoreceptors. Stem Cells, 2013. 31(6): p. 1149-59.
31. Lancaster, M.A., et al., Cerebral organoids model human brain development and microcephaly. Nature, 2013. 501(7467): p. 373-9.

2
Solving the Mystery of Memory

1 Luscher C, Malenka RC. Drug-evoked synaptic plasticity in addiction: from molecular changes to circuit remodeling. Neuron. 2011;69(4):650-63.
2 Flexner LB, Flexner JB, Roberts RB. Memory in mice analyzed with antibiotics. Antibiotics are useful to study stages of memory and to indicate molecular events which sustain memory. Science. 1967;155(768):1377-83.

3 Goelet P, Castellucci VF, Schacher S, Kandel ER. The long and the short of long-term memory--a molecular framework. Nature. 1986;322(6078):419-22.

4 Linzer DI, Nathans D. Growth-related changes in specific mRNAs of cultured mouse cells. Proceedings of the National Academy of Sciences of the United States of America. 1983;80(14):4271-5.

5 Lau LF, Nathans D. Identification of a set of genes expressed during the G0/G1 transition of cultured mouse cells. Embo J. 1985;4(12):3145-51.

6 Brakeman PR, Lanahan AA, O'Brien R, Roche K, Barnes CA, Huganir RL, et al. Homer: a protein that selectively binds metabotropic glutamate receptors. Nature. 1997;386(6622):284-8.

7 Beneken J, Tu JC, Xiao B, Nuriya M, Yuan JP, Worley PF, et al. Structure of the Homer EVH1 domain-peptide complex reveals a new twist in polyproline recognition. Neuron. 2000;26(1):143-54.

8 Frey U, Morris RG. Synaptic tagging and long-term potentiation. Nature. 1997;385(6616):533-6.

9 Orlando LR, Ayala R, Kett LR, Curley AA, Duffner J, Bragg DC, et al. Phosphorylation of the homer-binding domain of group I metabotropic glutamate receptors by cyclin-dependent kinase 5. Journal of neurochemistry. 2009;110(2):557-69.

10 Park JM, Hu JH, Milshteyn A, Zhang PW, Moore CG, Park S, et al. A prolyl-isomerase mediates dopamine-dependent plasticity and cocaine motor sensitization. Cell. 2013;154(3):637-50.

11 Hu JH, Park JM, Park S, Xiao B, Dehoff MH, Kim S, et al. Homeostatic scaling requires group I mGluR activation mediated by Homer1a. Neuron. 2010;68(6):1128-42. PMCID: 3013614.

12 Ango F, Prezeau L, Muller T, Tu JC, Xiao B, Worley PF, et al. Agonist-independent activation of metabotropic glutamate receptors by the intracellular protein Homer. Nature. 2001;411(6840):962-5.

13 Gray CM, Maldonado PE, Wilson M, McNaughton B. Tetrodes markedly improve the reliability and yield of multiple single-unit isolation from multi-unit recordings in cat striate cortex. J Neurosci Methods. 1995;63(1-2):43-54.

14 Wilson MA, McNaughton BL. Reactivation of hippocampal ensemble memories during sleep. Science. 1994;265(5172):676-9.

15 Foster DJ, Wilson MA. Reverse replay of behavioural sequences in hippocampal place cells during the awake state. Nature. 2006;440(7084):680-3.

16 Pfeiffer BE, Foster DJ. Hippocampal place-cell sequences depict future paths to remembered goals. Nature. 2013;497(7447):74-9.

17 Guzowski JF, McNaughton BL, Barnes CA, Worley PF. Environment-specific expression of the immediate-early gene Arc in hippocampal neuronal ensembles. Nature neuroscience. 1999;2(12):1120-4.

18 Vazdarjanova A, McNaughton BL, Barnes CA, Worley PF, Guzowski JF. Experience-dependent coincident expression of the effector immediate-early genes arc and Homer 1a in hippocampal and neocortical neuronal networks. The Journal of neuroscience : the official journal of the Society for Neuroscience. 2002;22(23):10067-71.

19 Guzowski JF, Timlin JA, Roysam B, McNaughton BL, Worley PF, Barnes CA. Mapping behaviorally relevant neural circuits with immediate-early gene expression. Current opinion in neurobiology. 2005;15(5):599-606.

20 Ramirez-Amaya V, Vazdarjanova A, Mikhael D, Rosi S, Worley PF, Barnes CA. Spatial exploration-induced Arc mRNA and protein expression: evidence for selective, network-specific reactivation. The Journal of neuroscience : the official journal of the Society for Neuroscience. 2005;25(7):1761-8.
21 Marrone DF, Schaner MJ, McNaughton BL, Worley PF, Barnes CA. Immediate-early gene expression at rest recapitulates recent experience. The Journal of neuroscience : the official journal of the Society for Neuroscience. 2008;28(5):1030-3.
22 Gavornik JP, Shuler MG, Loewenstein Y, Bear MF, Shouval HZ. Learning reward timing in cortex through reward dependent expression of synaptic plasticity. Proceedings of the National Academy of Sciences of the United States of America. 2009;106(16):6826-31. PMCID: 2672535.

3

Mapping Your Every Move

1 Olesen J, Leonardi M. The burden of brain diseases in Europe. Eur J Neurol. Sept 2003;10(5):471-7.
2 Olesen J, Gustavsson A, Svenssond M, Wittchene H-U and Jönsson B on behalf of the CDBE2010 study group and the European Brain Council. The economic cost of brain disorders in Europe. European Journal of Neurology 2012; 19: 155–162.
3 Tolman, EC. Cognitive maps in rats and men. Psychol. Rev. 1948; 55: 189-208.
4 O'Keefe J, Dostrovsky J. The hippocampus as a spatial map. Preliminary evidence from unit activity in the freely-moving rat. Brain Res. 1971; 34 (1): 171–5.
5 O'Keefe J, Nadel L. The Hippocampus as a Cognitive Map. 1978. Oxford: Clarendon Press.
6 Taube JS, Muller RU, Ranck JB, Jr. Head-direction cells recorded from the postsubiculum in freely moving rats. I. Description and quantitative analysis. The Journal of neuroscience: the official journal of the Society for Neuroscience. 1990; 10: 420-435
7 Taube JS, Muller RU, Ranck JB Jr. Head-direction cells recorded from the postsubiculum in freely moving rats. II. Effects of environmental manipulations. The Journal of neuroscience: the official journal of the Society for Neuroscience. 1990; 10: 436-447.
8 Hafting T, Fyhn M, Molden S, Moser M-B, Moser EI. Microstructure of a spatial map in the entorhinal cortex. Nature. 2005; 436, 801–806.
9 Savelli F, Yoganarasimha D, Knierim JJ. Influence of boundary removal on the spatial representations of the medial entorhinal cortex. Hippocampus. 2008; 18:1270-1282.
10 Solstad T, Boccara CN, Kropff E, Moser M-B, and Moser EI. Representation of geometric borders in the entorhinal cortex. Science. 2008; 322:1865–1868.
11 Stensola H, Stensola T, Solstad T, Frøland K, Moser M-B, and Moser EI. The entorhinal grid map is discretized. Nature. 2012; 492: 72–78.
12 Zhang SJ, Ye J, Miao C, Tsao A, Cerniauskas I, Ledergerber D, Moser M-B, Moser EI. Optogenetic dissection of the entorhinal-hippocampal functional connectivity. Science. 2013; 340:1232627

4
Equal ≠ The Same

1 http://www.cbsnews.com/news/sex-matters-drugs-can-affect-sexes-differently/
2 Levine, S. Sex differences in the brain. Scientific American 1966; 214, 84–90.
3 Beery, A and Zucker, I. Sex bias in neuroscience and biomedical research. Neuroscience & Biobehavioral Reviews, 2011; 35, 565–572.
4 Eagly, A et al., Feminism and Psychology- Analysis of a Half-Century of Research on Women and Gender, American Psychologist, 2012; 67, 211-230.
5 http://townhall.com/columnists/johnstossel/2014/03/12/war-on-women-n1807016
6 Cahill, L. Why Sex Matters for Neuroscience. Nature Neuroscience Reviews, 2006; 7, 477-484.
7 Jazin, E and Cahill, L. Sex Differences in Molecular Neuroscience: From Drosophila to Humans. Nature Neuroscience Reviews, 2010; 11: 9-17.
8 Hines, M., Brain Gender, 2004, Oxford Univ Press.
9 Ingalhalikar, M et a., Sex differences in the structural connectome of the human brain, PNAS (USA), 2014; 111, 823-828.
10 Cahill, L. Fundamental sex difference in human brain architecture. PNAS (USA), 2014, 111, 577-578.
11 Jahanshad, N et al., Sex Differences in the human brain connectome: 4-Tesla angular resolution diffusion imaging (HARDI) tractography in 234 adult twins,
12 Biomedical Imaging: From Nano to Macro, IEEE International Symposium, 2011, 939-943.
13 Duarte-Carvajalino, J et al., Hierarchical topological network analysis of anatomical human brain connectivity and differences related to sex and kinship, NeuroImage, 2012; 59, 3784–3804.
14 Gong, G and Evans, A., Brain Connectivity: Gender makes a difference, The Neurowcientist, 2011, 17, 575-591.
15 Cribbs, D et al., Extensive innate immune gene activation accompanies brain aging, increasing vulnerability to cognitive decline and neurodegeneration: a microarray study, Journal of Neuroinflammation, 2012; 9, 179.
16 Fine, C., Is There Neurosexism in Functional Neuroimaging Investigations of Sex Differences? Neuroethics, 2012, DOI 10.1007/s12152-012-9169-1.
17 Jordan-Young, R. Brain Storm: The Flaws in the Science of Sex Differences, 2010, Harvard University Press.
18 Hyde, J, The gender similarities hypothesis. American Psychologist, 2005; 60, 581-592.
19 Burnett, S., Sex-related differences in spatial ability: Are they trivial? American Psychologist, 1986, 41, 1012 1013.
20 Del Giudice, M., The Distance Between Mars and Venus: Measuring Global Sex Differences in Personality, PLOS ONE, 2012; 7, 1-8.
21 Carothers, B. and Reis, H. Men and Women Are From Earth: Examining the Latent Structure of Gender. Journal of Personality and Social Psychology. 2012 Advance online publication. doi: 10.1037/a0030437
22 Lippa, R., Sex Differences in Personality Traits and Gender-Related Occupational Preferences across 53 Nations: Testing Evolutionary and Social-Environmental Theories Arch Sex Behav (2010) 39:619–636.

23 De Vries, G , Sex Differences in Adult and Developing Brains: Compensation, Compensation, Compensation, Endocrinology 2004, 145, 1063-1068
24 Joel, D Genetic-gonadal-genitals sex (3G-sex) and the misconception of brain and gender, or, why 3G-males and 3G-females have intersex brain and intersex gender. Biology of Sex Differences 2012, 3:27.
25 http://www.haaretz.com/news/features/.premium-1.576554
26 Wu, H. et al., Cellular Resolution Maps of X Chromosome Inactivation: Implications for Neural Development, Function, and Disease, Neuron, 2014; 81, 103-119.
27 Nadaf, S et al., Activity map of the tammar X chromosome shows that marsupial X inactivation is incomplete and escape is stochastic, Genome Biology, 2010; 11, 1-18.
28 Kopsida, E. et al., The role of the Y chromosone in brain function, Open Neuroendocrinol J, 2009 ; 2: 20–30. doi:10.2174/1876528900902010020.
29 Pinker, S. The Sexual Paradox, Scribner, NY, 2008, p.44.
30 Fine, C. et al. Plasticity, plasticity, plasticity. . . and the rigid problem of sex, Trends in Cognitive Sciences November 2013, Vol. 17, No. 11.
31 Eliot, L., Pink Brain, Blue Brain: How Small Differences Grow Into Troublesome Gaps -- And What We Can Do About It, 2009; HMH Publishing.
32 Udry, J. Biological Limits of Gender Construction, American Sociological Review, 2000; 65, 443-457.
33 Pinker, S., The Blank Slate: The Modern Denial of Human Nature, 2002, Penquin group.
34 Darwin, C. "Descent of Man and Selection in Relation to Sex", 2nd Ed, John Murray, London, 1875, Preface to the Second Edition, page vi.
35 Schenck-Gustafsson, K et al., Handbook of Clinical Gender Medicine, Karger Press, Basel, 2012.

5
Rich Man, Poor Man

1. Asimov, I., In the Game of Energy and Thermodynamics You Can't Even Break Even: Journal of Smithsonian Institue. 1970, June.
2. Toga, A.W., Thompson, P.M., and Sowell, E.R., Mapping brain maturation. Trends in Neurosciences, 2006. 29(3): p. 148-159.
3. Purves, D. and Lichtman, J.W., Principles of Neural Development. 1985, Sunderland, MA: Sinauer Associates, Incorporated. 433.
4. Shore, R., Rethinking the Brain: New Insights into Early Development. 1997, New York, NY: Families and Work Institute. 92.
5. Huttenlocher, P.R. and Dabholkar, A.S., Regional differences in synaptogenesis in human cerebral cortex. Journal of Comparative Neurology, 1997. 387: p. 167-178.
6. Neville, H.J. and Bavelier, D., Neural organization and plasticity of language. Current Opinion in Neurobiology, 1998. 8(2): p. 254-8.
7. White-Schwoch, T., Carr, K.W., Anderson, S., Strait, D.L., and Kraus, N., Older adults benefit from music training early in life: biological evidence for long-term training-driven plasticity. The Journal of neuroscience : the official journal of the Society for Neuroscience, 2013. 33(45): p. 17667-74.

8. National Center for Children in Poverty. 2014; Available from: http://www.nccp.org/.
9. Brooks-Gunn, J. and Duncan, G.J., The effects of poverty on children. The Future of Children, 1997. 7(2): p. 55-71.
10. Feinstein, L., Inequality in the early cognitive development of British children in the 1970 cohort. Economica, 2003. 70(277): p. 73-97.
11. Noble, K.G., McCandliss, B.D., and Farah, M.J., Socioeconomic gradients predict individual differences in neurocognitive abilities. Developmental Science, 2007. 10(4): p. 464-80.
12. Noble, K.G., Norman, M.F., and Farah, M.J., Neurocognitive correlates of socioeconomic status in kindergarten children. Developmental Science, 2005. 8(1): p. 74-87.
13. Farah, M.J., Shera, D.M., Savage, J.H., Betancourt, L., Giannetta, J.M., Brodsky, N.L., Malmud, E.K., and Hurt, H., Childhood poverty: Specific associations with neurocognitive development. Brain Research, 2006. 1: p. 166-74.
14. Fernald, L.C.H., Weber, A., Galasso, E., and Ratsifandrihamanana, L., Socioeconomic gradients and child development in a very low income population: Evidence from Madagascar. Developmental Science, 2011. 14(4): p. 832-847.
15. Brito, N.H., Engelhardt, L.E., Mack, L., Nail, E.J., Barr, R., Fifer, W.P., Elliott, A.J., and Noble, K.G., Socioeconomic Status is Associated with Language and Memory Development in the First Two Years, in International Society for Infant Studies. 2014: Berlin.
16. Tomalski, P., Moore, D.G., Ribeiro, H., Axelsson, E.L., Murphy, E., Karmiloff-Smith, A., Johnson, M.H., and Kushnerenko, E., Socioeconomic status and functional brain development – associations in early infancy. Developmental Science, 2013: p. n/a-n/a.
17. Benasich, A.A., Gou, Z., Choudhury, N., and Harris, K.D., Early cognitive and language skills are linked to resting frontal gamma power across the first 3 years. Behavioural Brain Research, 2008. 195(2): p. 215-222.
18. Gou, Z., Choudhury, N., and Benasich, A.A., Resting frontal gamma power at 16, 24 and 36 months predicts individual differences in language and cognition at 4 and 5 years. Behavioural Brain Research, 2011. 220(2): p. 263-270.
19. Noble, K.G., Houston, S.M., Kan, E., and Sowell, E.R., Neural correlates of socioeconomic status in the developing human brain. Developmental Science, 2012. 15(4): p. 516-527.
20. Hart, B. and Risley, T., Meaningful Differences in the Everyday Experience of Young American Children. 1995, Baltimore, MD: Brookes.
21. Kuhl, P.K., Brain Mechanisms in Early Language Acquisition. Neuron, 2010. 67(5): p. 713-727.
22. Kuhl, P.K., Is speech learning 'gated' by the social brain? Developmental Science, 2007. 10(1): p. 110-20.
23. Hanson, J.L., Chandra, A., Wolfe, B.L., and Pollak, S.D., Association between income and the hippocampus. PLoS ONE, 2011. 6(5): p. e18712.
24. Jednoróg, K., Altarelli, I., Monzalvo, K., Fluss, J., Dubois, J., Billard, C., Dehaene-Lambertz, G., and Ramus, F., The influence of socioeconomic status on children's brain structure. PLoS ONE, 2012. 7(8): p. e42486.
25. Noble, K.G., Grieve, S.M., Korgaonkar, M.S., Engelhardt, L.E., Griffith, E.Y., Williams, L.M., and Brickman, A.M., Hippocampal volume varies with educa-

tional attainment across the life-span. Frontiers in Human Neuroscience, 2012. 6(Article 307).

26. Staff, R.T., Murray, A.D., Ahearn, T.S., Mustafa, N., Fox, H.C., and Whalley, L.J., Childhood socioeconomic status and adult brain size: Childhood socioeconomic status influences adult hippocampal size. Annals of Neurology, 2012. 71(5): p. 653-660.
27. Champagne, D.L., Bagot, R.C., van Hasselt, F., Ramakers, G., Meaney, M.J., de Kloet, E.R., Jo, x00Eb, ls, M., and Krugers, H., Maternal care and hippocampal plasticity: Evidence for experience-dependent structural plasticity, altered synaptic functioning, and differential responsiveness to glucocorticoids and stress. Journal of Neuroscience, 2008. 28(23): p. 6037-45.
28. Tottenham, N. and Sheridan, M., A review of adversity, the amygdala, and the hippocampus: A consideration of developmental timing. Frontiers in Human Neuroscience, 2010. 3: p. 1-18.
29. McEwen, B.S. and Gianaros, P.J., Central role of the brain in stress and adaptation: Links to socioeconomic status, health, and disease. Annals of the New York Academy of Sciences, 2010. 1186(1): p. 190-222.
30. Buss, C., Lord, C., Wadiwalla, M., Hellhammer, D.H., Lupien, S.J., Meaney, M.J., and Pruessner, J.C., Maternal care modulates the relationship between prenatal risk and hippocampal volume in women but not in men. Journal of Neuroscience, 2007. 27(10): p. 2592-2595.
31. Mezzacappa, E., Alerting, orienting, and executive attention: Developmental properties and sociodemographic correlates in an epidemiological sample of young, urban children. Child Development, 2004. 75(5): p. 1373-86.
32. Blair, C. and Raver, C.C., Child development in the context of adversity: Experiential canalization of brain and behavior. American Psychologist, 2012. 67(4): p. 309-318.
33. Raver, C.C., Emotions matter: Making the case for the role of young children's emotional development for early school readiness. 2002, Harris School of Public Policy Studies, University of Chicago, Working Papers.
34. Raver, C.C., Garner, P.W., and Smith-Donald, R., The roles of emotion regulation and emotion knowledge for children's academic readiness: Are the links causal?, in School Readiness and the Transition to Kindergarten in the Era of Accountability. 2007, Paul H. Brookes Publishing: Baltimore, MD. p. 121-147.
35. Noble, K.G., Korgaonkar, M.S., Grieve, S.M., and Brickman, A.M., Higher Education is an Age-Independent Predictor of White Matter Integrity and Cognitive Control in Late Adolescence. Developmental Science, 2013.
36. Gianaros, P.J., Horenstein, J.A., Cohen, S., Matthews, K.A., Brown, S.M., Flory, J.D., Critchley, H.D., Manuck, S.B., and Hariri, A.R., Perigenual anterior cingulate morphology covaries with perceived social standing. Social Cognitive & Affective Neuroscience, 2007. 2(3): p. 161-73.
37. Lawson, G.M., Duda, J.T., Avants, B.B., Wu, J., and Farah, M.J., Associations between children's socioeconomic status and prefrontal cortical thickness. Developmental Science, 2013: p. n/a-n/a.
38. Luby, J., Belden, A., Botteron, K., Marrus, N., Harms, M.P., Babb, C., Nishino, T., and Barch, D., The Effects of Poverty on Childhood Brain Development: The Mediating Effect of Caregiving and Stressful Life Events. JAMA Pediatr, 2013.
39. Kim, P., Evans, G.W., Angstadt, M., Ho, S.S., Sripada, C.S., Swain, J.E., Liberzon,

I., and Phan, K.L., Effects of childhood poverty and chronic stress on emotion regulatory brain function in adulthood. Proceedings of the National Academy of Sciences of the United States of America, 2013. 110(46): p. 18442-7.

40. Stevens, C., Lauinger, B., and Neville, H., Differences in the neural mechanisms of selective attention in children from different socioeconomic backgrounds: An event-related brain potential study. Developmental Science, 2009. 12(4): p. 634-46.
41. Gee, D.G., Gabard-Durnam, L.J., Flannery, J., Goff, B., Humphreys, K.L., Telzer, E.H., Hare, T.A., Bookheimer, S.Y., and Tottenham, N., Early developmental emergence of human amygdala–prefrontal connectivity after maternal deprivation. Proceedings of the National Academy of Sciences, 2013. 110(39): p. 15638-15643.
42. Blair, C., Granger, D.A., Willoughby, M., Mills-Koonce, R., Cox, M., Greenberg, M.T., Kivlighan, K.T., Fortunato, C.K., and the, F.L.P.I., Salivary cortisol mediates effects of poverty and parenting on executive functions in early childhood. Child Development, 2011. 82(6): p. 1970-1984.
43. Raver, C.C., Jones, S.M., Li-Grining, C., Zhai, F., Bub, K., and Pressler, E., CSRP's impact on low-income preschoolers' preacademic skills: Self-regulation as a mediating mechanism. Child Development, 2011. 82(1): p. 362-378.
44. Weiland, C. and Yoshikawa, H., Impacts of a prekindergarten program on children's mathematics, language, literacy, executive function, and emotional skills. Child Development, 2013: p. ePub ahead of print.
45. Kalil, A., Ryan, R., and Corey, M., Diverging destinies: maternal education and the developmental gradient in time with children. Demography, 2012. 49(4): p. 1361-83.
46. Muennig, P., Schweinhart, L., Montie, J., and Neidell, M., Effects of a prekindergarten educational intervention on adult health: 37-year follow-up results of a randomized controlled trial. Journal Information, 2009. 99(8).
47. Campbell, F., Conti, G., Heckman, J.J., Moon, S.H., Pinto, R., Pungello, E., and Pan, Y., Early childhood investments substantially boost adult health. Science, 2014. 343(6178): p. 1478-85.
48. Waldfogel, J. and Washbrook, E., Early Years Policy. Child Development Research, 2011. 2011.
49. Fernald, A. How Talking to Children Nurtures Language Development Across SES and Culture. in American Association for the Advancement of Science. 2014. Chicago.
50. Kitzman, H.J., Olds, D.L., Cole, R.E., Hanks, C.A., Anson, E.A., Arcoleo, K.J., Luckey, D.W., Knudtson, M.D., Henderson, C.R., and Holmberg, J.R., Enduring effects of prenatal and infancy home visiting by nurses on children: follow-up of a randomized trial among children at age 12 years. Archives of Pediatrics & Adolescent Medicine, 2010. 164(5): p. 412-418.
51. Furstenberg, F.F., The Challenges of Finding Causal Links Between Family Educational Practices and Schooling Outcomes., in Whither Opportunity?: Rising Inequality, Schools, and Children's Life Chances, G.J. Duncan and R.J. Murnane, Editors. 2011, Russell Sage Foundation: New York. p. 465-482.
52. Dahl, G.B. and Lochner, L., The impact of family income on child achievement: Evidence from the earned income tax credit. The American Economic Review, 2012. 102(5): p. 1927-1956.

53. Ziol-Guest, K.M., Duncan, G.J., Kalil, A., and Boyce, W.T., Early childhood poverty, immune-mediated disease processes, and adult productivity. Proceedings of the National Academy of Sciences, 2012. 109(Supplement 2): p. 17289-17293.

6
One of a Kind

1. Steptoe, A., Wardle, J. & Marmot, M. Positive affect and health-related neuroendocrine, cardiovascular, and inflammatory processes. Proc. Natl. Acad. Sci. U. S. A. 102, 6508–12 (2005).
2. Fraga, M. F. et al. Epigenetic differences arise during the lifetime of monozygotic twins. Proc. Natl. Acad. Sci. U. S. A. 102, 10604–9 (2005).
3. Fox, A. S., Shelton, S. E., Oakes, T. R., Davidson, R. J. & Kalin, N. H. Trait-like brain activity during adolescence predicts anxious temperament in primates. PLoS One 3, e2570 (2008).
4. Oler, J. A. et al. Amygdalar and hippocampal substrates of anxious temperament differ in their heritability. Nature 466, 864–8 (2010).
5. Schaefer, S. M. et al. Purpose in life predicts better emotional recovery from negative stimuli. PLoS One 8, e80329 (2013).
6. Javaras, K. N. et al. Conscientiousness predicts greater recovery from negative emotion. Emotion 12, 875–81 (2012).
7. Schuyler, B. S. et al. Temporal dynamics of emotional responding: amygdala recovery predicts emotional traits. Soc. Cogn. Affect. Neurosci. (2012). doi:10.1093/scan/nss131
8. McMakin, D. L., Siegle, G. J. & Shirk, S. R. Positive Affect Stimulation and Sustainment (PASS) Module for Depressed Mood: A preliminary investigation of treatment-related effects. Cognit. Ther. Res. 35, 217–226 (2011).
9. Pickut, B. a et al. Mindfulness based intervention in Parkinson's disease leads to structural brain changes on MRI: a randomized controlled longitudinal trial. Clin. Neurol. Neurosurg. 115, 2419–25 (2013).
10. Weng, H.Y. et al. Compassion Training Alters Altruism and Neural Responses to Suffering. Psychol. Sci. 24, 1171–80 (2013).
11. LeDoux, J. Rethinking the emotional brain. Neuron 73, 653–76 (2012).
12. Birn, R. M. et al. Evolutionarily-conserved prefrontal-amygdalar dysfunction in early-life anxiety. Mol. Psychiatry (2014).
13. Schaefer, S. M. et al. Purpose in life predicts better emotional recovery from negative stimuli. PLoS One 8, e80329 (2013).
14. Javaras, K. N. et al. Conscientiousness predicts greater recovery from negative emotion. Emotion 12, 875–81 (2012).
15. McMakin, D. L., Siegle, G. J. & Shirk, S. R. Positive Affect Stimulation and Sustainment (PASS) Module for Depressed Mood: A preliminary investigation of treatment-related effects. Cognit. Ther. Res. 35, 217–226 (2011).
16. Heller, A. S. et al. Relationships Between Changes in Sustained Fronto-Striatal Connectivity and Positive Affect in Major Depression Resulting From Antidepressant Treatment. Am. J. Psychiatry 170, 197–206 (2013).
17. Heller, A. S. et al. Sustained ventral striatal activity predicts eudaimonic well-being and cortisol output. Psychol. Sci. 24, 2191–2200 (2013).

18. Steptoe, A., Dockray, S. & Wardle, J. Positive affect and psychobiological processes relevant to health. J. Pers. 77, 1747–76 (2009).
19. Davidson, R. J. & McEwen, B. S. Social influences on neuroplasticity: stress and interventions to promote well-being. Nat. Neurosci. 15, 689–95 (2012).
20. Hölzel, B. K. et al. Stress reduction correlates with structural changes in the amygdala. Soc. Cogn. Affect. Neurosci. 5, 11–7 (2010).
21. Pickut, B. a et al. Mindfulness based intervention in Parkinson's disease leads to structural brain changes on MRI: a randomized controlled longitudinal trial. Clin. Neurol. Neurosurg. 115, 2419–25 (2013).
22. Weng, H.Y. et al. Compassion Training Alters Altruism and Neural Responses to Suffering. Psychol. Sci. 24, 1171–80 (2013).

7
The Age Gauge

1. McGrath JJ, Petersen L, Agerbo E, Mors O, Mortensen P, Pedersen C. A comprehensive assessment of parental age and psychiatric disorders. JAMA Psychiatry. 2014;71(3):301-309.
2. D'Onofrio BM, Rickert ME, Frans EM, et al. Paternal age at childbearing and offspring psychiatric and academic morbidity. JAMA Psychiatry. 2014.
3. Miettunen J, Suvisaari J, Haukka J, Isohanni M. Use of register data for psychiatric epidemiology in the Nordic countries. In: Tsuang MT, Tohen M, Jones P, eds. Textbook of Psychiatric Epidemiology. 3rd ed. Chichester, UK: Wiley; 2011:117-131.
4. Byrne N, Regan C, Howard L. Administrative registers in psychiatric research: a systematic review of validity studies. Acta Psychiatr Scand. 2005;112:409-414.
5. Gaziano JM. The Evolution of Population Science. JAMA: The Journal of the American Medical Association. 2010;304(20):2288-2289.
6. Weissman MM, Brown AS, Talati A. Translational Epidemiology in Psychiatry: Linking Population to Clinical and Basic Sciences. Archives of General Psychiatry. June 1, 2011 2011;68(6):600-608.
7. Hiatt RA. Invited Commentary: The Epicenter of Translational Science. Am J Epidemiol. September 1, 2010 2010;172(5):525-527.
8. Khoury MJ, Gwinn M, Ioannidis JPA. The Emergence of Translational Epidemiology: From Scientific Discovery to Population Health Impact. Am J Epidemiol. September 1, 2010 2010;172(5):517-524.
9. D'Onofrio BM, Lahey BB, Turkheimer E, Lichtenstein P. The critical need for family-based, quasi-experimental research in integrating genetic and social science research. American Journal of Public Health. 2013;103:S46-S55.
10. Reichenberg A, Gross R, Weiser M, et al. Advancing paternal age and autism. Arch Gen Psychiatry. Sep 2006;63(9):1026-1032.
11. Hultman CM, Sandin S, Levine SZ, Lichtenstein P, Reichenberg A. Advancing paternal age and risk of autism: new evidence from a population-based study and a meta-analysis of epidemiological studies. Mol Psychiatry. 2011;16:1203-1212.
12. Lundstrom S, Haworth CM, Carlstrom E, et al. Trajectories leading to autism spectrum disorders are affected by paternal age: findings from two nationally

representative twin studies. Journal of child psychology and psychiatry, and allied disciplines. Jul 2010;51(7):850-856.

13. Miller B, Messias E, Miettunen J, et al. Meta-analysis of Paternal Age and Schizophrenia Risk in Male Versus Female Offspring. Schizophr Bull. September 1, 2011 2011;37(5):1039-1047.
14. Malaspina D, Harlap S, Fennig S, et al. Advancing paternal age and the risk of schizophrenia. Arch Gen Psychiatry. Apr 2001;58(4):361-367.
15. Frans EM, Sandin S, Reichenberg A, Lichtenstein P, Långström N, Hultman CM. Advancing Paternal Age and Bipolar Disorder. Archives of General Psychiatry. 2008;65:1034-1040.
16. Menezes PR, Lewis G, Rasmussen F, et al. Paternal and maternal ages at conception and risk of bipolar affective disorder in their offspring. Psychological Medicine. 2010;40(03):477-485.
17. Puleo CM, Reichenberg A, Smith CJ, Kryzak LA, Silverman JM. Do autism-related personality traits explain higher paternal age in autism? Mol Psychiatry. 2008;13:243-244.
18. Petersen L, Mortensen PB, Pedersen CB. Paternal age at birth of first child and risk of schizophrenia. American Journal of Psychiatry. 2011;168:82-88.
19. Granville-Grossman KL. Paternal age and schizophrenia. British Journal of Psychiatry. 1966;112:899-905.
20. Lahey BB, D'Onofrio BM. All in the family: Comparing siblings to test causal hypotheses regarding environmental influences on behavior. Current Directions in Psychological Science. 2010;19:319-323.
21. Susser E, Eide MG, Begg M. Invited Commentary: The Use of Sibship Studies to Detect Familial Confounding. Am J Epidemiol. September 1, 2010 2010;172(5):537-539.
22. Crow JF. Spontaneous mutation in man. Mutation Research/Reviews in Mutation Research. 1999;437(1):5-9.
23. Kong A, Frigge ML, Masson G, et al. Rate of de novo mutations and the importance of father's age to disease risk. Nature. Aug 23 2012;488(7412):471-475.
24. Veltman JA, Brunner HG. De novo mutations in human genetic disease. Nature Reviews Genetics. 2012;13:565-575.
25. Neale BM, Kou Y, Liu L, et al. Patterns and rates of exonic de novo mutations in autism spectrum disorders. Nature. May 10 2012;485(7397):242-245.
26. O'Roak BJ, Vives L, Girirajan S, et al. Sporadic autism exomes reveal a highly interconnected protein network of de novo mutations. Nature. May 10 2012;485(7397):246-250.
27. Sanders SJ, Murtha MT, Gupta AR, et al. De novo mutations revealed by whole-exome sequencing are strongly associated with autism. Nature. May 10 2012;485(7397):237-241.
28. Harden KP. Genetic influences on adolescent sexual behavior: Why genes matter for environmentally oriented researchers. Psychological Bulletin. 2014;140(2):434-465.
29. Frans EM. High paternal age and risk of psychiatric disorders in offspring. Stockholm, Sweden: Karolinska Institutet; 2013.
30. Powell B, Steelman LC, Carini RM. Advancing Age, Advantaged Youth: Parental Age and the Transmission of Resources to Children. Soc Forces. 2006;84(3):1359-1390.

31. Mills M, Rindfuss RR, McDonald P, te Velde E, on behalf of the ESHRE Reproduction Society Task Force. Why do people postpone parenthood? Reasons and social policy incentives. Hum Reprod Update. 2011;17(6):848-860.
32. Academy of Medical Sciences Working Group. Identifying the environmental causes of disease: How should we decide what to believe and when to take action? London: Academy of Medical Sciences; 2007.

8
The Time of Your Life

1. Reppert, S.M. & Weaver, D.R. Molecular analysis of mammalian circadian rhythms. Annu Rev Physiol 63, 647-76 (2001).
2. Schibler, U., and Sassone-Corsi, P. (2002). A web of circadian pacemakers. Cell 111, 919-922.
3. Dodd AN1, Salathia N, Hall A, Kévei E, Tóth R, Nagy F, Hibberd JM, Millar AJ, Webb AA. (2005) Plant circadian clocks increase photosynthesis, growth, survival and competitive advantage. Science 309: 630-33.
4. Hunt T. and Sassone-Corsi P. (2007) Riding the tandem: circadian clocks and the cell cycle. Cell 129: 461-64.
5. Welsh, D.K., Takahashi, J.S. & Kay, S.A. Suprachiasmatic nucleus: cell autonomy and network properties. Annu Rev Physiol 72, 551-77 (2010).
6. Whitmore D, Foulkes NS, Strähle U, Sassone-Corsi P. (1998) Zebrafish clock rhythmic expression reveals independent peripheral circadian oscillators. Nat Neurosci. 1: 701-7.
7. Giebultowicz, J. M. and Hege, D. M. (1997). Circadian clock in Malpighian tubules. Nature 386, 664
8. Plautz, J. D., Kaneko, M., Hall, J. C. and Kay, S. A. (1997). Independent photoreceptive circadian clocks throughout Drosophila. Science 278, 1632 -1635.
9. Yamazaki, S., Numano, R., Abe, M., Hida, A., Takahashi, R., Ueda, M., Block, G. D., Sakaki, Y., Menaker, M. and Tei, H. (2000). Resetting central and peripheral circadian oscillators in transgenic rats. Science 288, 682 -685.
10. Pando MP, Morse D, Cermakian N, Sassone-Corsi P. (2002) Phenotypic rescue of a peripheral clock genetic defect via SCN hierarchical dominance. Cell 110: 107-17.
11. Balsalobre, A., Damiola, F. and Schibler, U. (1998). A serum shock induces circadian gene expression in mammalian tissue culture cells. Cell 93, 929 -937.
12. Konopka RJ and Benzer S. (1971) Clock mutants of Drosophila melanogaster. Proc Natl Acad Sci U S A. 68: 2112-6.
13. Masri, S., and Sassone-Corsi, P. (2012). Plasticity and specificity of the circadian epigenome. Nature Neuroscience 13, 1324-1329.
14. Katada, S., and Sassone-Corsi, P. (2010). The histone methyltransferase MLL1 permits the oscillation of circadian gene expression. Nature structural & molecular biology 17, 1414-1421.
15. Nakahata, Y., et al. 2008. The NAD+-dependent deacetylase SIRT1 modulates CLOCK-mediated chromatin remodeling and circadian control. Cell. 134: 329-340.
16. Asher, G., et al. 2008. SIRT1 regulates circadian clock gene expression through PER2 deacetylation. Cell. 134: 317-328.

17. Nakahata, Y., Sahar, S., Astarita, G., Kaluzova, M., and Sassone-Corsi, P. (2009). Circadian Control of the NAD+ Salvage Pathway by CLOCK-SIRT1. Science 324: 654-657
18. Ramsey, K.M., Yoshino, J., Brace, C.S., Abrassart, D., Kobayashi, Y., Marcheva, B., Hong, H.K., Chong, J.L., Buhr, E.D., Lee, C., et al. (2009). Circadian Clock Feedback Cycle Through NAMPT-Mediated NAD+ Biosynthesis. Science 324: 651-654
19. Eckel-Mahan KL, Patel VR, de Mateo S, Orozco-Solis R, Ceglia NJ, Sahar S, Dilag-Penilla SA, Dyar KA, Baldi P, Sassone-Corsi P. (2013) Reprogramming of the circadian clock by nutritional challenge. Cell 155: 1464-78
20. Masri S. Rigor P., Cervantes M., Ceglia N., Sebastian C., Xiao C., Roqueta-Rivera M., Deng C., Osborne T., Mostoslavsky R., Baldi P. and Sassone-Corsi P. (2014) Partitioning Circadian Transcription by SIRT6 leads to Segregated Control of Cellular Metabolism. Cell (in press)

9
Brain-to-Brain Interfaces

1 Carmena, J.M., M.A. Lebedev, R.E. Crist, J.E. O'Doherty, D.M. Santucci, D.F. Dimitrov, P.G. Patil, C.S. Henriquez, and M.A. Nicolelis, Learning to control a brain-machine interface for reaching and grasping by primates. PLoS Biol, 2003. 1(2): p. E42.261882.
2 Nicolelis, M., Beyond boundaries: the new neuroscience of connecting brains with machines--and how it will change our lives. 1st ed2011, New York: Times Books/Henry Holt and Co. 353 p.
3 O'Doherty, J.E., M.A. Lebedev, P.J. Ifft, K.Z. Zhuang, S. Shokur, H. Bleuler, and M.A. Nicolelis, Active tactile exploration using a brain-machine-brain interface. Nature, 2011. 479(7372): p. 228-31.3236080.
4 Pais-Vieira, M., M. Lebedev, C. Kunicki, J. Wang, and M.A. Nicolelis, A brain-to-brain interface for real-time sharing of sensorimotor information. Sci Rep, 2013. 3: p. 1319.3584574.

10
With A Little Help from Our Friends

1. Harlow HF (1958) The nature of love. Amer Psychologist 13, 673-685.
2. Harlow HF, Dodsworth RO, Harlow MK (1965) Total social isolation in monkeys. Proc Natl Acad Sci 54, 90-7.
3. Kanari K, Kikusui T, Takeuchi Y, Mori Y (2005) Multidimensional structure of anxiety-related behavior in early-weaned rats. Behav Brain Res 156:45-52.
4. Darwin C (1871) The Descent of Man.
5. James W (1890) The Principles of Psychology.
6. Chartrand TL, Bargh JA (1999) The chameleon effect: the perception-behavior link and social interaction. J Pers Soc Psychol 76:893-910.
7. Gregory SW (1990) Analysis of fundamental frequency reveals covariation in interview partners' speech. J Nonverbal Behavior 14:237-51.
8. Niedenthal PM (2007) Embodying emotion. Science 316:1002-5.

9. Laird JD, Bresler C (1992) The process of emotional experience: A self-perception theory. In: Clark MS (Ed), Emotion. Review of personality and social psychology, No. 13 (pp. 213-234). Thousand Oaks, CA, US: Sage Publications, Inc.
10. Preston SD, de Waal FB (2002) Empathy: Its ultimate and proximate bases. Behav Brain Sci 25:1-20.
11. Batson CD, Fultz J, Schoenrade PA (1987) Distress and empathy: two qualitatively distinct vicarious emotions with different motivational consequences. J Pers 55:19-39.
12. de Waal FB (2008) Putting the altruism back into altruism: the evolution of empathy. Annu Rev Psychol 59:279-300.
13. Warneken F, Hare B, Melis AP, Hanus D, Tomasello M (2007) Spontaneous altruism by chimpanzees and young children. PLoS Biol 5:e184.
14. Decety J (2011) The neuroevolution of empathy. Ann N Y Acad Sci 1231:35-45.
15. Panksepp J, Panksepp JB (2013) Toward a cross-species understanding of empathy. Trends Neurosci 36:489-96.
16. Jeon D, Kim S, Chetana M, Jo D, Ruley HE, Lin SY, Rabah D, Kinet JP, Shin HS (2010) Observational fear learning involves affective pain system and Cav1.2 Ca2+ channels in ACC. Nat Neurosci 13:482-8.
17. Langford DJ, Crager SE, Shehzad Z, Smith SB, Sotocinal SG, Levenstadt JS, Chanda ML, Levitin DJ, Mogil JS (2006) Social modulation of pain as evidence for empathy in mice. Science 312:1967-70.
18. Ben-Ami Bartal I, Decety J, Mason P (2011) Empathy and pro-social behavior in rats. Science 334:1427-30.
19. Ben-Ami Bartal I, Rodgers DA, Bernardez Sarria MS, Decety J, Mason P (2014) Pro-social behavior in rats is modulated by social experience. eLife 3:e01385.
20. Tajfel H (1970) Experiments in intergroup discrimination. Sci Amer 223:96-102.
21. Silberberg A, Allouch C, Sandfort S, Kearns D, Karpel H, Slotnick B (2014) Desire for social contact, not empathy, may explain "rescue" behavior in rats. Anim Cogn 17:609-18.
22. Clay Z, de Waal FB (2013) Development of socio-emotional competence in bonobos. Proc Natl Acad Sci 110: 18121-18126.
23. Stoltenberg SF, Christ CC, Carlo G (2013) Afraid to help: social anxiety partially mediates the association between 5-HTTLPR triallelic genotype and prosocial behavior. Soc Neurosci 8: 400-406.
24. Decety J, Yang CY, Cheng Y (2010) Physicians down-regulate their pain empathy response: an event-related brain potential study. Neuroimage 50:1676-82.
25. Jackson PL, Rainville P, Decety J (2006) To what extent do we share the pain of others? Insight from the neural bases of pain empathy. Pain 125:5-9.
26. Cheng Y, Lin CP, Liu HL, Hsu YY, Lim KE, Hung D, Decety J (2007) Expertise modulates the perception of pain in others. Curr Biol 17:1708-13.

11
The Brain-Games Conundrum

1. SharpBrains (2013). Executive summary: Infographic on the digital brain health market 2012-2020. Retrieved September 23, 2014, from http://sharpbrains.com/executive-summary/

2. Zhang, P. (n.d.). Infographic: celebrating 50 million Lumosity members. Retrieved September 23, 2014, from http://blog.lumosity.com/50millioninfographic/
3. Nintendo (n.d.). Top selling software units. Retrieved September 23, 2014, from http://www.nintendo.co.jp/ir/en/sales/software/ds.html
4. Taatgen, N.A. (2013). The nature and transfer of cognitive skills. Psychological Review, 120(3), 439-471.
5. Thorndike, E. L. (1906). Principles of teaching. New York: Seiler.
6. Ericsson, K. A., & Lehmann, A. C. (1996). Expert and exceptional performance: Evidence of maximal adaptation to task constraints. Annual Review of Psychology, 47(1), 273-305.
7. Logan, G. D. (1988). Toward an instance theory of automatization. Psychological Review, 95(4), 492-527.
8. Gopher, D., Weil, M., & Siegel, D. (1989). Practice under changing priorities: An approach to the training of complex skills. Acta Psychologica, 71, 147-177.
9. Schmidt, R. A., & Bjork, R. A. (1992). New conceptualizations of practice: Common principles in three paradigms suggest new concepts for training. Psychological Science, 3(4), 207-217.
10. Dahlin, E., Neely, A. S., Larsson, A., Bäckman, L., & Nyberg, L. (2008). Transfer of learning after updating training mediated by the striatum. Science, 320(5882), 1510-1512.
11. Bavelier, D., Green, C. S., Pouget, A., & Schrater, P. (2012). Brain plasticity through the life span: learning to learn and action video games. Annual Review of Neuroscience, 35, 391-416.
12. Tennstedt, S. L., & Unverzagt, F. W. (2013). The ACTIVE Study Study Overview and Major Findings. Journal of Aging and Health, 25(8 suppl), 3S-20S.
13. Owen, A. M., Hampshire, A., Grahn, J. A., Stenton, R., Dajani, S., Burns, A. S., ... & Ballard, C. G. (2010). Putting brain training to the test. Nature, 465(7299), 775-778.
14. Mayas, J., Parmentier, F. B., Andrés, P., & Ballesteros, S. (2014). Plasticity of attentional functions in older adults after non-action video game training: a randomized controlled trial. PloS One, 9(3), e92269.
15. Boot, W. R., Simons, D. J., Stothart, C., & Stutts, C. (2013). The pervasive problem with placebos in psychology why active control groups are not sufficient to rule out placebo effects. Perspectives on Psychological Science, 8(4), 445-454.
16. Baniqued, P., Kranz, M.B., Voss, M., Lee, H., Cosman, J.D., Severson, J. & Kramer, A.F. (2014). Cognitive training with casual video games: Points to consider. Frontiers in Psychology, January, vol 4, article 1010, 1-19.
17. Jaeggi, S. M., Buschkuehl, M., Jonides, J., & Perrig, W. J. (2008). Improving fluid intelligence with training on working memory. Proceedings of the National Academy of Sciences, 105(19), 6829-6833.
18. Shipstead, Z., Redick, T. S., & Engle, R. W. (2012). Is working memory training effective? Psychological Bulletin, 138(4), 628-654.
19. Harrison, T. L., Shipstead, Z., Hicks, K. L., Hambrick, D. Z., Redick, T. S., & Engle, R. W. (2013). Working memory training may increase working memory capacity but not fluid intelligence. Psychological Science, 0956797613492984.
20. Redick, T. S., Shipstead, Z., Harrison, T. L., Hicks, K. L., Fried, D. E., Hambrick, D. Z., ... & Engle, R. W. (2013). No evidence of intelligence improvement after

working memory training: a randomized, placebo-controlled study. Journal of Experimental Psychology: General, 142(2), 359-379.

21. Thompson, T., Waskom, M.L., Garel., K.L.A., Cardenas-Iniguez, C., Reynolds, G.O., Winter, R., Chang, P., Pollard, K., Lala, N., Alvarez, G.A. & Gabrieli, J.D.E. (2013). Failure of working memory training to enhance cognition or intelligence. PLOS One, 8 (5), 1-15.
22. Au, J., Sheehan, E., Tsai, N., Duncan, G. J., Buschkuehl, M., & Jaeggi, S. M. (2014, e-pub ahead of print). Improving fluid intelligence with training on working memory: a meta-analysis. Psychonomic Bulletin & Review, 1-12.
23. Jaeggi, S. M., Buschkuehl, M., Shah, P., & Jonides, J. (2014). The role of individual differences in cognitive training and transfer. Memory & Cognition, 42(3), 464-480.
24. Green, C. S., & Bavelier, D. (2003). Action video game modifies visual selective attention. Nature, 423(6939), 534-537.
25. Powers, K. L., Brooks, P. J., Aldrich, N. J., Palladino, M. A., & Alfieri, L. (2013). Effects of video-game play on information processing: a meta-analytic investigation. Psychonomic Bulletin & Review, 20(6), 1055-1079.
26. Basak, C., Boot, W. R., Voss, M. W., & Kramer, A. F. (2008). Can training in a real-time strategy video game attenuate cognitive decline in older adults? Psychology and Aging, 23(4), 765-777.
27. Boot, W. R., Kramer, A. F., Simons, D. J., Fabiani, M., & Gratton, G. (2008). The effects of video game playing on attention, memory, and executive control. Acta Psychologica, 129(3), 387-398.
28. van Ravenzwaaij, D., Boekel, W., Forstmann, B. U., Ratcliff, R., & Wagenmakers, E. J. (2014, e-pub ahead of press). Action video games do not improve the speed of information processing in simple perceptual tasks. Journal of Experimental Psychology: General.
29. Boot, W. R., Blakely, D. P., & Simons, D. J. (2011). Do action video games improve perception and cognition? Frontiers in Psychology, 2.
30. Kristjánsson, Á. (2013). The case for causal influences of action videogame play upon vision and attention. Attention, Perception, & Psychophysics, 75(4), 667-672.
31. Voss M, Vivar C, Kramer AF, van Praag H (2013) Bridging animal and human models of exercise-induced brain plasticity. Trends in Cognitive Sciences, 17, 525-544.
32. Carlson, M. C., Erickson, K. I., Kramer, A. F., Voss, M. W., Bolea, N., Mielke, M., ... & Fried, L. P. (2009). Evidence for neurocognitive plasticity in at-risk older adults: the experience corps program. The Journals of Gerontology Series A: Biological Sciences and Medical Sciences, glp117.
33. Park, D. C., Lodi-Smith, J., Drew, L., Haber, S., Hebrank, A., Bischof, G. N., & Aamodt, W. (2014). The Impact of Sustained Engagement on Cognitive Function in Older Adults The Synapse Project. Psychological science, 25(1), 103-112.
34. Stine-Morrow, E.A.L., Payne, B.R., Roberts, B.W., Kramer, A.F., Morrow, D.G., Payne. L., Hill, P.L., Jackson, J.J., Gao, X., Noh, S.R., Janke, M.C. & Parisi, J.M. (in press). Training versus engagement to cognitive enrichment with aging. Psychology & Aging.
35. Boot, W. R., Champion, M., Blakely, D. P., Wright, T., Souders, D. J., & Charness, N. (2013). Video games as a means to reduce age-related cognitive decline: attitudes, compliance, and effectiveness. Frontiers in Psychology, 4.

36. Hertzog, C., Kramer, A. F., Wilson, R. S., & Lindenberger, U. (2008). Enrichment effects on adult cognitive development can the functional capacity of older adults be preserved and enhanced?. Psychological Science in the Public Interest, 9(1), 1-65.
37. McAuley, E., Mullen, S.P., Szabo, A.N., White, S.M., Wojcicki, T.R., Mailey, E.L., Gothe, N.P., Olson, E.A., Voss, M., Erickson, K., Prakash, R. & Kramer, A.F. (2011). Self-regulatory processes and exercise adherence in older adults: Executive function and self-efficacy effects. American Journal of Preventive Medicine, 41(3), 284-290.
38. Rebok, G.W., Langbaum, J. B., Jones, R. N., Gross, A. L., Parisi, J. M., Spira, A. P., ... & Brandt, J. (2013). Memory Training in the ACTIVE Study How Much is Needed and Who Benefits? Journal of Aging and Health, 25(8 suppl), 21S-42S.
39. Stothart, C. R., Simons, D. J., Boot, W. R., & Kramer, A. F. (in press). Is the effect of aerobic exercise on cognition a placebo effect? PLOS One, 1-16 pages.
40. McDaniel, M. A., Binder, E., Bugg, J. M., Waldum, E., Dufault, C., Meyer, A., ...Kudelka, C. (2014). Effects of cognitive training with and without aerobic exercise on cognitively demanding everyday activities. Psychology and Aging, 29 (3), 717-730.
41. Prakash, R.S., Voss, M.W., Erickson, K.I. & Kramer, A.F. (in press). Moving towards a healthier brain and mind. Annual Review of Psychology.

12
You Say You Want a Revolution?

1. Heath, Chip, and Heath Dan, Decisive: How to Make Better Choices in Life and Work. New York: Crown Publishing, 2013.
2. Oshinsky, David, M., Polio: An American Story. New York: Oxford University Press, 2006.
3. Lu P, Blesch A, Graham L, Wang Y, Samara R, Banos K, Haringer V, Havton L, Weishaupt N, Bennet D, Fouad K, Tuszynski MH. Motor axonal regeneration after partial and complete spinal cord transection. J. Neuroscience 2012. 32(24): 8208-18.
4. Liu K, Lu Y, Lee JK, Samara R, Willenberg R, Sears-Kraxberger I, Tedeschi A, Park KK, Jin D, Cai B, Xu B, Connolly L, Steward O, Zheng B, He Z. PTEN deletion enhances the regenerative ability of adult corticospinal neurons. Nature Neuroscience. 2010. 13(9):1075-81.
5. Hebb, Donald O., The Organization of Behavior: A Neuropsychological Theory. New York: John Wiley & Sons, Inc., 1949.
6. Harold, Eve, Stem Cell Wars: Inside Stories from the Front Lines. New York: Palgrave Macmillan, 2006.
7. Weintraub, Karen, The Trials of Stem Cell Therapy: Stem Cells: Plenty of Hope, but Halting Progress. New York: New York Times, September 16, 2014.
8. CBS News (Sept. 24, 2014)
9. The Atlantic (Oct., 2014): http://www.theatlantic.com/features/archive/2014/09/why-i-hope-to-die-at-75/379329/

Index

C

D

E

F

G

H

I

J

K

L

M

N

O

P

Q

R

U

V

W

Y

Z